The twentieth century has seen biology come of age as a conceptual and quantitative science. Biochemistry, cytology, and genetics have been unified into a common framework at the molecular level. However, cellular activity and development are regulated not by the interplay of molecules alone, but by interactions of molecules organized in complex arrays, subunits, and organelles. Emphasis on organization is, therefore, of increasing importance.

So it is too, at the other end of the scale. Organismic and population biology are developing new rigor in such established and emerging disciplines as ecology, evolution, and ethology, but again the accent is on interactions between individuals, populations, and societies. Advances in comparative biochemistry and physiology have given new impetus to studies of animal and plant diversity. Microbiology has matured, with the world of viruses and procaryotes assuming a major position. New connections are being forged with other disciplines outside biology—chemistry, physics, mathematics, geology, anthropology, and psychology provide us with new theories and experimental tools while at the same time are themselves being enriched by the biologists' new insights into the world of life. The need to preserve a habitable environment for future generations should encourage increasing collaboration between diverse disciplines.

The purpose of the Modern Biology Series is to introduce the college biology student—as well as the gifted secondary student and all interested readers—both to the concepts unifying the fields within biology and to the diversity that makes each field unique.

Since the series is open-ended, it will provide a greater number and variety of topics than can be accommodated in many introductory courses. It remains the responsibility of the instructor to make his selection, to arrange it in a logical order, and to develop a framework into which the individual units can best be fitted.

New titles will be added to the present list as new fields emerge, existing fields advance, and new authors of ability and talent appear. Only thus, we feel, can we keep pace with the explosion of knowledge in Modern Biology.

James D. Ebert
Howard A. Schneiderman

Modern Biology Series ***Consulting Editors***

James D. Ebert
Marine Biological Laboratory

Howard A. Schneiderman
University of California, Irvine

Berns	***Cells***
Delevoryas	***Plant Diversification,*** *second edition*
Ebert/Sussex	***Interacting Systems in Development,*** *second edition*
Ehrenfeld	***Biological Conservation***
Fingerman	***Animal Diversity,*** *second edition*
Griffin/Novick	***Animal Structure and Function,*** *second edition*
Levine	***Genetics,*** *second edition*
Novikoff/Holtzman	***Cells and Organelles,*** *second edition*
Odum	***Ecology,*** *second edition*
Ray	***The Living Plant,*** *second edition*
Savage	***Evolution,*** *third edition*
Sistrom	***Microbial Life,*** *second edition*
Van der Kloot	***Behavior***

Plant Diversification

Second Edition

Theodore Delevoryas

The University of Texas at Austin

Holt, Rinehart and Winston

New York Chicago San Francisco Atlanta
Dallas Montreal Toronto London Sydney

Library of Congress Cataloging in Publication Data
Delevoryas, Theodore, 1929–
Plant diversification.

(Modern biology series)
Includes index.
1. Botany—Morphology. 2. Plants—Evolution.
3. Botany—Variation. I. Title.
QK641.D4 1977 581.3′8 76-30858
ISBN 0-03-080133-8

Printed in the United States of America
0 9 8 7 090 1 2 3 4 5 6 7 8 9

To Nan

Preface

I have long felt that it is virtually impossible to present a survey of the entire plant kingdom as part of an introductory botany or biology course. The superficial treatment so often given to the plant kingdom in such a course leaves the student with a vague notion of all the groups and with sound knowledge of practically none. For that reason, I have not attempted in this slim volume to discuss all the known groups of plants. My plan is, instead, to select certain topics of special evolutionary interest and to develop them more fully than is possible in a survey of all the plants. My selection of important evolutionary sequences may not have been entirely objective, nor the same as what other writers consider important, but I am quite confident that a good number of the topics selected represent highlights in the evolution of the plant kingdom.

An evolutionary treatment of the plant kingdom becomes more meaningful when the groups are identified as diverse manifestations of a certain basic body plan and life history. Therefore, throughout the book there has been an attempt to stress homologies among the various groups and to interpret evolutionary changes as modifications of the basic, generalized plan of these organisms. The pitfalls in this method are evident, and it is necessary to recognize similarities in form in many cases as convergence among the groups. A consideration of convergence as a result of a similar response of independent plant groups to environmental and other factors adds considerable interest to the subject of plant evolution.

The main emphasis of this book is on *form* and the evolution of plant form, rather than plant groups. However, form cannot be discussed on this introductory level in the abstract. In order for new forms to originate in the plant kingdom, they must have something to evolve from, and actual structures in actual plants are utilized as examples. Those organisms which are named and described are not included for their own sakes, but are merely illustrative of the particular evolutionary theme under consideration.

Also included, where applicable, is a consideration of the fossil record and its contribution to our knowledge of the changes that occurred in the plant kingdom through the ages. Paleobotanical information adds the important element of *time*, an appreciation of which is essential in any evolutionary study. Where possible, the discussion of form is related to its association with a particular function in the plant, and the role of the environment in shaping plant form is included.

A large part of this book is concerned with the evolution of flowering plants. One of the reasons for this is that the angiosperms constitute the most conspicuous part of our vegetable environment at the present time, and are the group with which most students have some acquaintance. An appreciation of this immediate part of our environment will be more lasting, especially among those who do not continue in the botanical sciences, than a knowledge of other groups, even though evolutionary problems in the latter can be just as fascinating as those among the angiosperms.

I hope it will be understood by the student that the evolutionary series presented throughout this book are not to be regarded as demonstrated facts. These series are, instead, logical conclusions held by many botanists based on observation and interpretation of the facts available to us. Additional information about plant structure, together with a reinterpretation of known facts, could lead to different hypotheses in the future. Indeed, advances in knowledge since the appearance of the first edition have led to a revision of some of the suggested evolutionary series presented in that edition. Evolutionary plant morphology is not a "closed book," and we can expect many new ideas as time goes on. It is quite possible—I sincerely hope probable—that readers will be stimulated toward reevaluating some of the facts and ultimately presenting new ideas that may lead to a more accurate picture of change in the plant kingdom.

This book has benefited greatly from criticisms and suggestions of a number of people who read the manuscript. I am most indebted to my friend and colleague Professor Harold C. Bold, whose help and advice were of inestimable value. Errors in fact or interpretation, however, should be blamed on the author.

Without the cooperation of many authors and publishers who granted permission to reproduce illustrations, this book could not have been possible.

Austin, Texas
December 1976 T. D.

Contents

Plant
Diversification

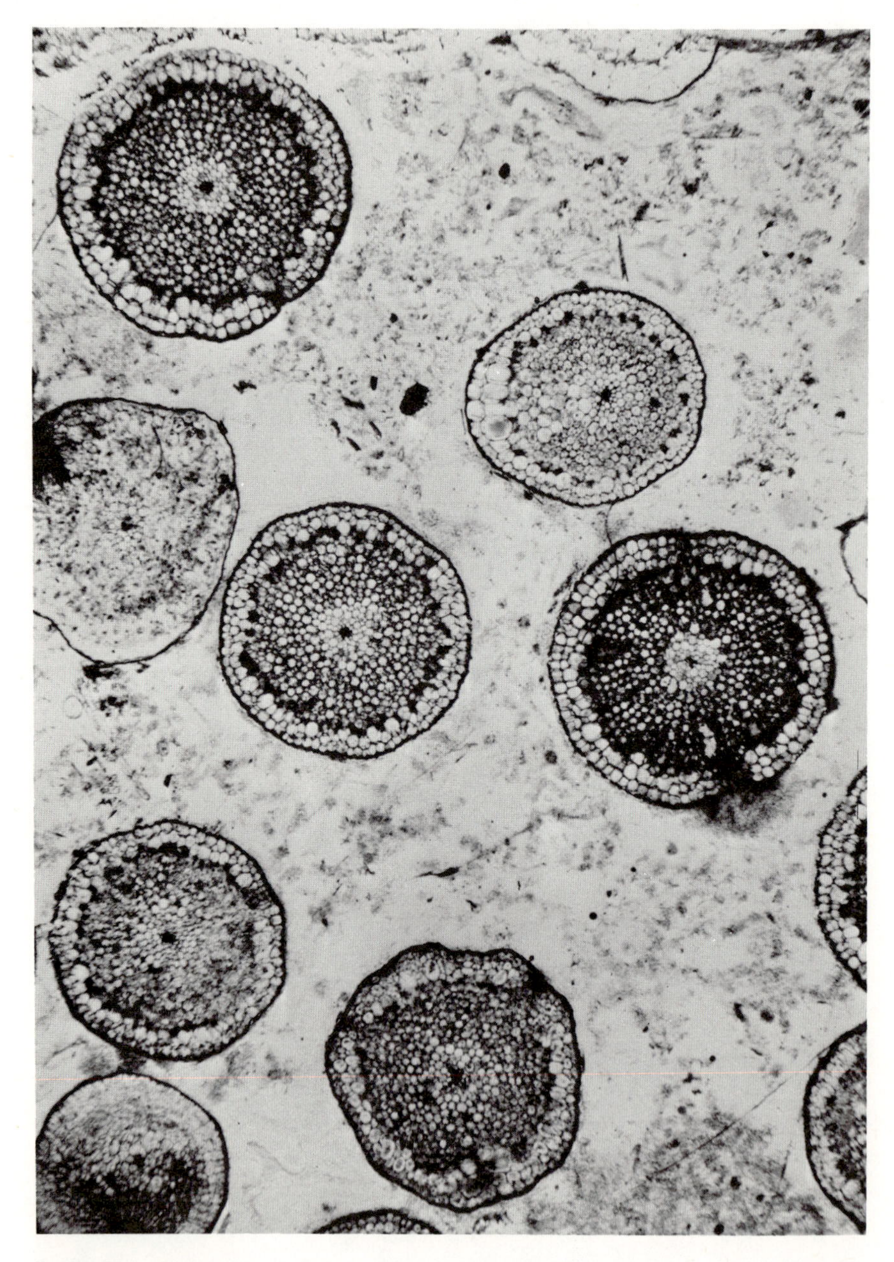

Frontispiece. Thin section of Devonian chert from Aberdeenshire, Scotland, showing transverse sections of stems of *Rhynia Gwynne-Vaughani*, an extremely simple vascular plant. × 16.

Introduction

Even to a nonbotanist it is apparent that a great variety of plants cover the earth's surface. A person is aware of trees in the forests, crops raised on farmlands, lawns and ornamentals around homes, curious green scums that appear and disappear in ponds, fungus growths, both fleshy and otherwise, delicate and filmy plants along moist brooksides, and a countless variety of other forms. Obviously, this diversification of plants did not appear suddenly on the earth, but is the result of long periods of evolutionary change—change in response to a number of factors. It is one of the purposes of this book to attempt to trace the beginnings of plant life on the earth, the modifications that occurred in the structure and functioning of these plants, the problems posed to these evolving plants in history and at the present time by an environment that itself is changing, and the various kinds of ways these problems were solved by some of the different groups of plants. This study involves a history of the plant kingdom as well as an attempt to explain the relationship between structure and function in both living and extinct plants.

DISTINGUISHING FEATURES OF PLANTS

Most people are quite aware of what plants are, and in most instances it is a simple matter to distinguish a plant from other kinds of living things. No one has difficulty in differentiating between an oak tree and an elephant, for example. On the other hand, there are certain groups of plants that are not so obviously different from some animal groups. In fact, some organisms are classified as plants by botanists and as animals by zoologists.

Often biologists institute another kingdom of living things, the Protista, which includes some of the simple organisms that seem to combine characteristics of both plants and animals. Although such a scheme has some usefulness, there are certain objections, the main one being that it is an artificial grouping of obviously unrelated organisms.

What are the main criteria by which plants can be distinguished from all other living creatures? Perhaps the most obvious characteristic of plants is the presence of chlorophyll, and the ability of plants to utilize this green pigment in trapping energy from the sun for food-synthesizing reactions. Most plants, we might say, are efficient and independent machines that provide their own food reserves. However, exceptions to this concept of plants come to mind immediately; fungi, of course, lack this green photosynthetic pigment and are dependent upon outside sources for nourishment.

Another characteristic that seems to be present in most plants is a stationary type of existence; plants are not generally thought of as motile organisms, while animals usually have the ability to get about from one place to another. Again, this characteristic of motile versus nonmotile habit is not always valid; many plantlike organisms have means of moving about in an aqueous environment. Furthermore, certain animals are anchored to one spot and spend their entire existence there.

The inability to distinguish all plants from all animals need not be a disturbing situation. While all higher plants are distinct from all higher animals, it is among the organisms, both plant and animal, that are considered to be primitive that the distinction becomes vaguer and the two kingdoms appear to merge. It is generally recognized by biologists that the two kingdoms began their independent courses of evolution from similar types of simple and primitive organisms. Earliest plants and animals were more like each other than are the more highly evolved, later forms of plants and animals. Also, parallelisms between the two groups of organisms are quite common among the simpler forms. Subsequent to the initial divergence of the two kingdoms (conceivably, there might have been more than one divergence), each line followed its own independent course. In many instances, both plants and animals met with the same kinds of problems in their environments. These problems were "solved" with different means by the two groups. As differences became more pronounced, the organisms in the two kingdoms responded to environmental challenges in even more conspicuously divergent ways. Thus, while plants and animals have much in common in their basic natures, there are also tremendous dissimilarities.

PLANT KINGDOM DIVISIONS

Because references will be made to various plant groups by name throughout the book, a short survey of the plant kingdom is presented in this initial chapter as a quick means of placing these groups in their proper places. Practically all botanists have their own ideas about the way in which the plant kingdom should be classified and of the possible relationships of the groups. The scheme presented here is not intended to be definitive (the element of subjectivity is hard to eliminate in the classifying of organisms); it is only one relatively simple way of grouping organisms to show their positions in relation to each other. Modern ideas concerning phylogeny of the plant kingdom are incorporated in this plan. Students may prefer to use some other scheme; the main part of the book is quite suitably adapted to a number of systems of classification. The following system of classification of the plant kingdom has been chosen.

***Blue-green algae* (Cyanochloronta)** Simple unicellular, colonial, or multicellular plants lacking membrane-bounded nuclei and with pigments not in organized bodies. Principal pigments are chlorophyll *a*, β-carotene, xanthophylls, phycoerythrin, and phycocyanins. There is no sexual reproduction or motile flagellated cells.

***Euglenophytes* (Euglenophycophyta)** Simple unicells or colonial motile organisms. Motility is effected by flagella. Pigments (chlorophylls *a* and *b*, β-carotene) in organized bodies called plastids. Food-storage products are carbohydrates related to starch and fats.

***Green algae* (Chlorophycophyta)** Unicellular, colonial, or multicellular green plants with generally simple structure. Principal pigments are chlorophylls *a* and *b*, carotenes, and xanthophylls contained within plastids. The food-storage product is generally starch. Reproduction is by asexual or sexual means.

***Golden algae* (Chrysophycophyta)** A group of simple plants paralleling the green algae in form. Pigments include chlorophyll *a* (other chlorophylls may be present, but not chlorophyll *b*), carotenes, and xanthophylls. The food-storage product is generally an oil or a carbohydrate called leucosin.

***Brown algae* (Phaeophycophyta)** Structurally the most complex algae, ranging from simple unicells to massive plant bodies. Pigments include chlorophylls *a* and *c*, β-carotene, and xanthophylls. Food stored as oil or a carbohydrate. Both sexual and asexual means of reproduction are present.

***Red algae* (Rhodophycophyta)** Most forms multicellular; pigment bodies contain chlorophylls *a* and *d*, carotene, a xanthophyll, and a reddish pigment, phycoerythrin. Sexual and asexual reproductive systems are present; there are no flagellated reproductive cells.

***Dinoflagellates* (Pyrrophycophyta)** Unicellular, flagellated organisms with principal pigments chlorophylls *a* and *c*, β-carotene, and xanthophylls. Cells are generally covered with a number of platelike structures.

***Bacteria* (Schizonta)** Extremely small, unicellular or filamentous chains. Mostly heterotrophic, with reproduction typically by cell division.

***Plasmodial slime molds* (Myxomycota)** Vegetative phase of the life history consisting of a naked, streaming mass of protoplasm. Specialized sporangia are produced at the time of spore formation.

***Cellular slime molds* (Acrasiomycota)** Vegetative phase consisting of amoebalike cells that aggregate into a streaming mass before fruiting.

***Chytrids* (Chytridiomycota)** Usually aquatic, some parasitic, colorless organisms producing motile reproductive cells, each with one posterior flagellum.

***Oomycetous fungi* (Oomycota)** Nonseptate, filamentous organisms producing motile reproductive cells, each with two dissimilar flagella. (This division includes the "water molds.")

***Zygomycetous fungi* (Zygomycota)** Usually nonseptate, filamentous organisms with nonmotile reproductive cells. Sexual reproduction involves fusion of contents of two equal cells.

***Sac fungi* (Ascomycota)** Some unicellular, most multicellular, with septations in filaments. Special spores develop within elongated sacs that are produced only after the process of sexual fusion.

***Club fungi* (Basidiomycota)** Multicellular fungi with special spores borne on the tip of a generally club-shaped structure formed some time following sexual fusion.

***Bryophytes* (Bryophyta)** Multicellular green plants with an alternation of haploid and diploid phases. The haploid, gamete-bearing phase is typically the more conspicuous. This division includes the mosses and liverworts; these are sometimes placed in separate divisions.

The following are divisions of vascular plants. These are highly organized green plants with specialized conducting tissue (wood and phloem) in the diploid, spore-producing phase.

***Rhyniophytes* (Rhyniophyta)** Extinct vascular plants constructed of dichotomously branched, naked axes with sporangia borne terminally.

***Zosterophyllophytes* (Zosterophyllophyta)** Extinct vascular plants with generally dichotomously branched axes, sometimes bearing spines, with sporangia borne along the sides of the axes.

***Trimerophytes* (Trimerophyta)** Extinct vascular plants consisting of generally leafless principal axes, with smaller branched laterals bearing terminal sporangia.

***Lycopods or club mosses* (Lycophyta)** Leafy plants with closely spaced simple leaves. Sporangia are borne on the upper sides of the leaves or in the angles between leaves and stem.

***Articulates or horsetails* (Sphenophyta)** Plants with leaves and branches generally borne in whorls. Sporangia are borne on specialized fertile branches that are often aggregated into cones.

***Ferns* (Pterophyta)** Plants with leaves that are typically large and often compound, with sporangia generally borne on the lower surfaces. Stems may be elongated, and either horizontal or erect.

***Progymnosperms* (Progymnospermophyta)** Extinct plants producing abundant secondary vascular tissues and often heterosporous. This division is considered to have been a precursor of the seed plants.

***Seed ferns* (Pteridospermophyta)** Plants with the general habit of ferns, but with seeds borne on the leaves.

***Cycads* (Cycadophyta)** Plants with generally woody stems and leaves superficially resembling those of palms. Seeds are borne on modified leaves that are typically aggregated into a cone.

***Cycadeoids* (Cycadeoidophyta)** Extinct plants superficially resembling cycads but with reproductive structures borne in complex, compact cones.

***Ginkgoes* (Ginkgophyta)** Trees with abundant wood and fan-shaped leaves. Seeds naked, borne on elongated stalks.

***Conifers* (Coniferophyta)** Trees with simple leaves that range from needlelike to flattened and broad. Seeds typically borne on the surfaces of woody scales aggregated into cones.

***Gnetophytes* (Gnetophyta)** Small group of unusual, naked-seeded plants that may represent more than one division.

***Flowering plants* (Anthophyta)** Flowering plants, with seeds typically borne enclosed within a structure called a carpel.

The categories of plants listed above include only the major divisions. Below "division" in descending magnitude, is "subdivision," followed by "class," "order," "family," "genus," and finally "species." Whenever possible, generalizations in this book will be based on the larger groups, but in discussions of form in plants it is often necessary to refer to smaller categories.

EARLIEST GEOLOGICAL EVIDENCE OF PLANT LIFE

Although the scope of this book is on the range of diversification of plants, it is interesting to go back in the earth's history to a time even before plants were on earth and to try to determine how life may have begun on the earth and what subsequent changes occurred in living things before they were recognizable as plants. It is generally assumed that the earth is about 4.5 billion years old and at the time of its formation was a large, gaseous aggregate of materials that gradually condensed. Gravitational force pulled the particles in toward the center, and the heaviest accumulated in the innermost regions. Around the central, more dense, and very hot part was an atmosphere in which elements that are now involved in living material were present. These elements are carbon (in methane gas), hydrogen, and nitrogen (in ammonia), in addition to water (in vapor form). As condensation progressed, water settled on earth in liquid form, and other elements accumulated in the ocean basins. Radiations from the sun provided the energy source for a number of kinds of chemical reactions, some of which resulted in the synthesis of organic compounds. Furthermore, lightning discharges could have played an important part in activating certain chemical reactions. Combinations of some of the simple amino acids thus produced into protein molecules represented a step toward the formation of living matter. Sugars and probably some other early kinds of organic compounds subsequently combined with phosphates (in solution as a result of chemical weathering of the earth's surface) and nitrogen bases and served as the necessary beginning of nucleic acids. Earliest forms of life undoubtedly synthesized new compounds utilizing materials in solution. To carry on reactions within the living material, energy needs to be utilized, and the source of this energy is the breakdown of products within the living being. Gradually, accumulation of carbon dioxide provided a different kind of atmosphere. This preceded the ability of simple organisms to synthesize their own organic food materials and eventually resulted in the process of photosynthesis. Without doubt, it may be stated that the world as it exists today is absolutely dependent upon this process.

Coal is a rock that is generally believed to have been derived from plant material. Most of the coal in the world represents remains of plant life that existed during the Carboniferous Period, some 250 to 350 million years ago. Other coals are known, however, that are quite a bit younger, as well as some older ones. Of great interest are coal deposits from the Precambrian Period, which are suggestive of the possible existence of plants in abundance during that part of the geologic past. Certain carbonaceous shales of the Precambrian Period also suggest the former presence of living organisms in the environment of deposition.

A problem of interest to laymen and biologists alike involves the first evidence of green, photosynthesizing organisms on earth. As time goes on, more and more evidence is accumulating to suggest that plants have been on earth for an extremely long time. Paleobotanists have been excavating plant remains in increasing abundance from rocks of Precambrian age; at one time it was believed by most paleontologists that no living organisms existed on the earth during that part of geologic time (about the first seven-eighths of geologic time). (Because numerous references will be made to various periods and eras of the geologic past, a timetable, Fig. 1-1, is included here for reference.)

The ultimate answer to the problem of plant origin will probably not be in the field of biology, but in the realm of paleogeochemistry. Chemically oriented geologists are able to discern evidence of organic compounds probably derived as a result of metabolic processes in plants and animals in past ages even when no actual visually observable structural remains are to be found. Such "chemical fossils" have been detected in rocks exceeding three billion years in age.

Many other indirect pieces of evidence attest to the great antiquity of the plant kingdom. For example, in many parts of the world there are massive deposits of limestone that have a peculiar concentric layering (Fig. 1-2A). We know that certain extant plants, blue-green algae in particular, are responsible for the accumulation of spherical or ellipsoidal nodules called "water biscuits," which are composed of concentrically added calcium carbonate deposits (Fig. 1-2B), and it has been suggested that some of these ancient, concentrically layered rocks, many of them Precambrian, could have had such an origin. Furthermore, other Precambrian carbonate deposits, often in the form of columnar structures with obvious layering (called stromatolites), are also probably algal in origin. Such stromatolites are being built up at the present time in the ocean off the west coast of Australia.

These are only a few examples of indirect evidence of living things, some of them presumably plant, known from some of the most ancient rocks. More certain and convincing, obviously, would be structurally preserved remains of these organisms that would give some idea of the structural nature of the earliest photosynthesizing plants. Fortunately, the physical processes involved in the precipitation of chert, a siliceous rock, allow the preservation of organisms that were living in the aqueous medium at that time. Structurally preserved organisms more than three

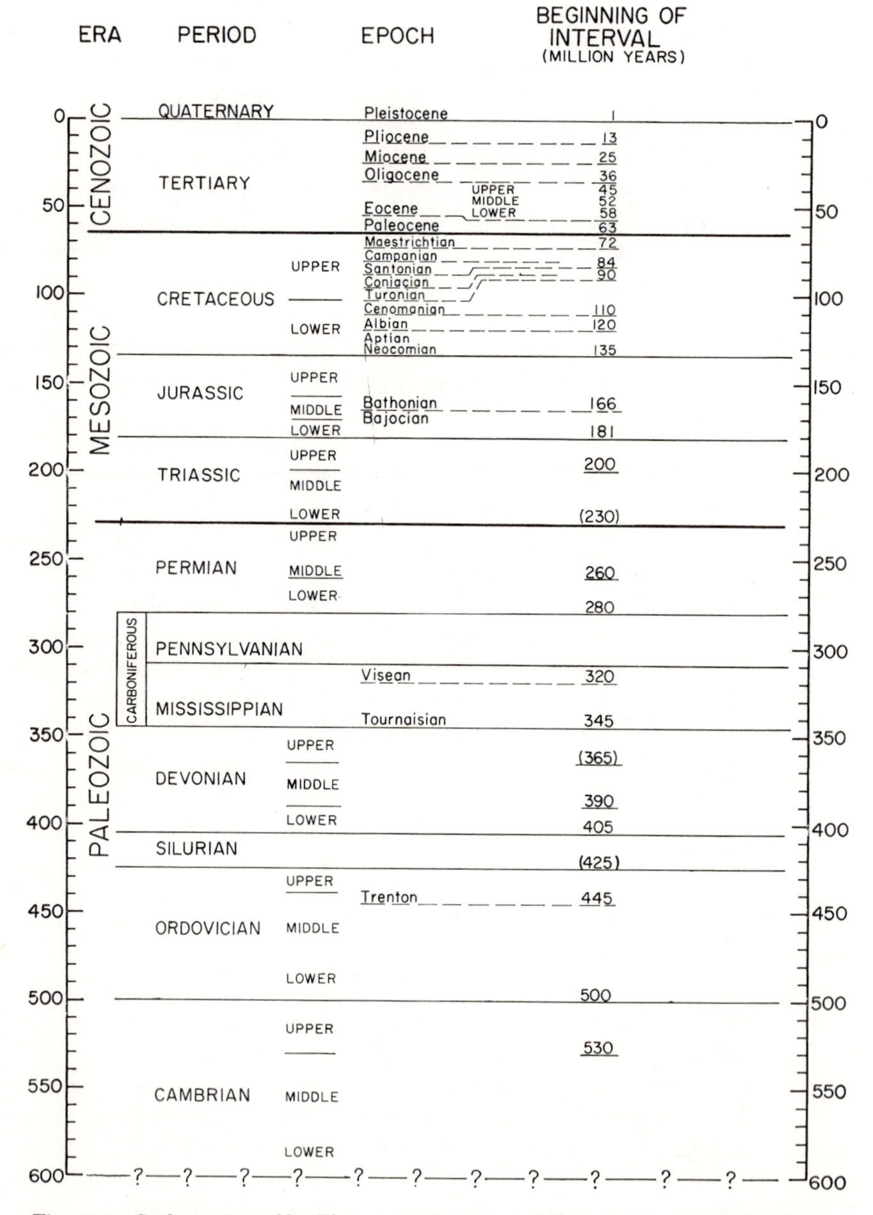

Fig. 1-1 *Geologic timetable. The principal portion of geologic time, the Precambrian Period, is below the bottom line on this scale. (From J. L. Kulp,* Science, *vol. 133, p. 1111, 1961. By permission.)*

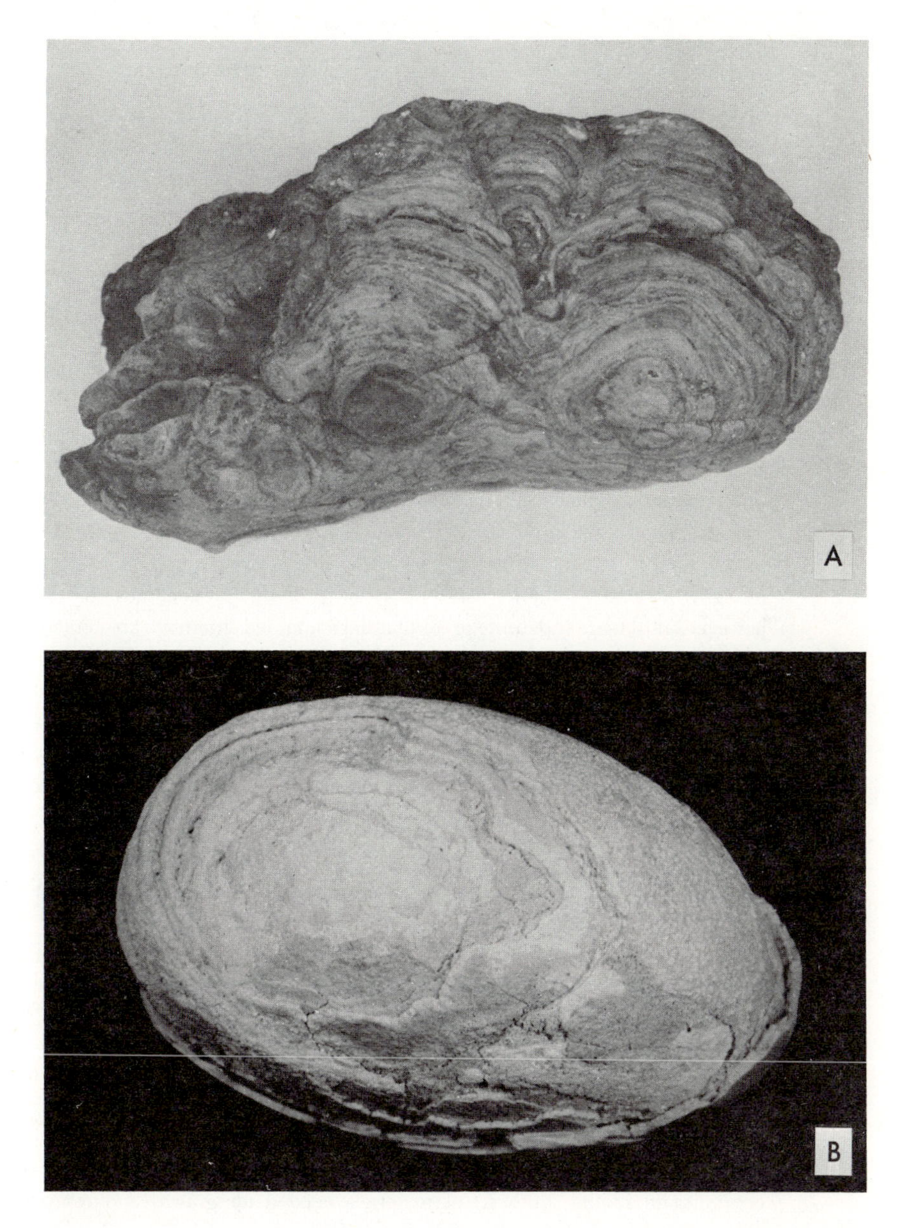

Fig. 1-2 (*A*) Cryptozoon, *a Precambrian limestone, possibly of blue-green algal origin, showing concentric layering.* $\times\frac{1}{2}$. (*B*) *"Water biscuit," a calcium carbonate deposit produced by blue-green algae in a Florida lake.* $\times 1\frac{1}{3}$.

billion years old have been observed in thin sections made of the chert. Most of the organisms are simple unicells that do not tell us much about the groups of known organisms to which they were most closely related, although unicells morphologically similar to bacteria are known from rocks about that old. In chert deposits almost two billion years old have been seen a greater diversity of organisms, some in an excellent state of preservation (Fig. 1-3). In these rocks are unicellular forms, filamentous and multicellular organisms, and filamentous, nonseptate structures. Probably the most frequently occurring forms are those having structural similarity to blue-green algae. Bacteria, too, are present in these sediments as well as a large group of fossils with unknown affinities. It is assumed that the organisms that lived in this part of the earth's history were primitive in the sense that they probably lacked organized nuclei. The suggestion that many may have been algae is made only on the basis of structural similarity, and it should be emphasized that nothing is known of the chemistry or the metabolism and reproductive processes of these organisms. It is even possible that certain of these fossillike remains may have been produced by chemical processes.

Fossil microorganisms from the Precambrian Period in Australia (dating to almost one billion years ago) have been discovered in an excellent state of preservation. Numerous blue-green algalike filaments are present as well as spherical unicells. Some of the latter have structures resembling nuclei and if, indeed, they are nuclei, their existence would serve as evidence that eucaryotic organisms appeared on earth at least a billion years before the present. Caution must be expressed, however, until there is definite proof that the contained structures are actually nuclei and not some result of the preservation process.

SIMPLEST FORMS OF PLANTS

Whether the first living unicellular (or subcellular) organisms were actually plants or animals is purely academic. Most likely they were neither. The first self-replicating organisms may have been nothing other than complex assemblages of organic molecules among which were included nucleic acids. Metabolism and replication probably occurred by a utilization of other compounds in the aqueous medium surrounding these primitive "organisms." Strictly speaking, then, the first mode of nutrition must have been a heterotrophic one. These organismlike beings, because of the nature of nucleic acid molecules, were able to reproduce similar types of individuals in a fairly consistent manner. From such forms, which may be considered heterotrophic, evolved even more complex organisms with a more elaborate and organized cell as the basis of structure. Some of these unicells, with more refined metabolic activities, could probably utilize simpler chemicals in their environment in the synthesis of materials needed for growth and reproduction.

Certain of the unicellular types, probably at various stages in the evolution of the plant kingdom, were able eventually to synthesize the complex organic compound, chlorophyll, which functions as a means for trapping the solar energy utilized

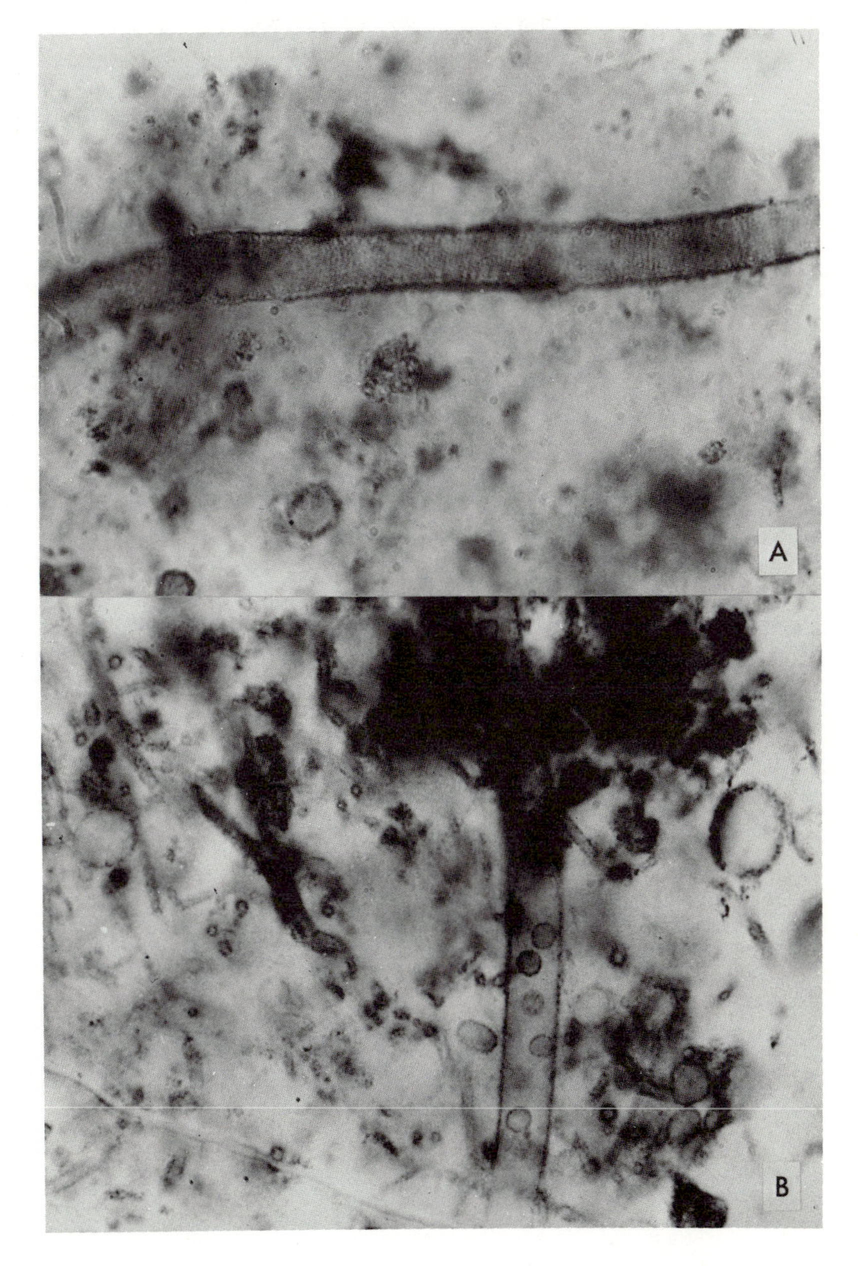

Fig. 1-3 *Sections of blue-green algalike fossils from the Precambrian Gunflint chert of southern Ontario.* (*A*) Animikiea septata. ×*1000.* (*B*) Entosphaeroides amplus. ×*1485.* (*Photographs courtesy of E. S. Barghoorn.*)

in the synthesizing of carbohydrates from carbon dioxide and water. The possibility that chlorophyll-bearing organisms could have evolved more than once in history has some basis in the fact that there are several kinds of chlorophylls in various groups of organisms; moreover, associated with these green pigments are many other kinds of pigments in varying combinations and proportions. Furthermore, the products of synthesis are quite different in the various groups of green aquatic organisms.

One interesting hypothesis that attempts to explain the origin of chloroplasts in plants suggests that plastids are green microorganisms that have come to live symbiotically within other plant cells. These organisms have lost their ability to live independently, but the fact that they seem to be able to reproduce within the host cells at the time of cell division is used as evidence that they may be individual organisms.

FURTHER READING

Barghoorn, E. S., "Origin of Life," in *Treatise on Marine Ecology and Paleoecology*, vol. 2. Geological Society of America, Monograph vol. 67 (1957), pp. 75–86.

Barghoorn, E. S., and S. A. Tyler, "Microorganisms from the Gunflint Chert," *Science*, vol. 147 (1965), pp. 563–577.

Schopf, J. W., "Microflora of the Bitter Springs Formation, Late Precambrian, Central Australia," *Journal of Paleontology*, vol. 42 (1968), pp. 651–688.

Schopf, J. W., "Precambrian Microorganisms and Evolutionary Events prior to the Origin of Vascular Plants," in *Major Evolutionary Events and the Geological Record of Plants. Biological Reviews of the Cambridge Philosophical Society*, vol. 45 (1970), pp. 319–352.

Wald, G., "The Origin of Life," *Scientific American*, vol. 191 (August 1954), pp. 44–53.

Some Evolutionary Trends in the Algae

Algae are chlorophyll-bearing, generally aquatic plants, ranging in form from extremely simple unicellular organisms to massive, complex types. They occupy a wide variety of habitats from pole to pole. Groups of algae show a number of conspicuous evolutionary trends, many of which involve changes from simplicity to progressive complexity.

UNICELLULAR ALGAE

Among the chlorophyll-bearing organisms that are considered to be the most primitive are the unicellular blue-green algae. These nonmotile algae have an extremely simple structure and method of reproduction. One group of blue-green algae, exemplified by the common genus *Gloeocapsa* (Fig. 2-1), has a

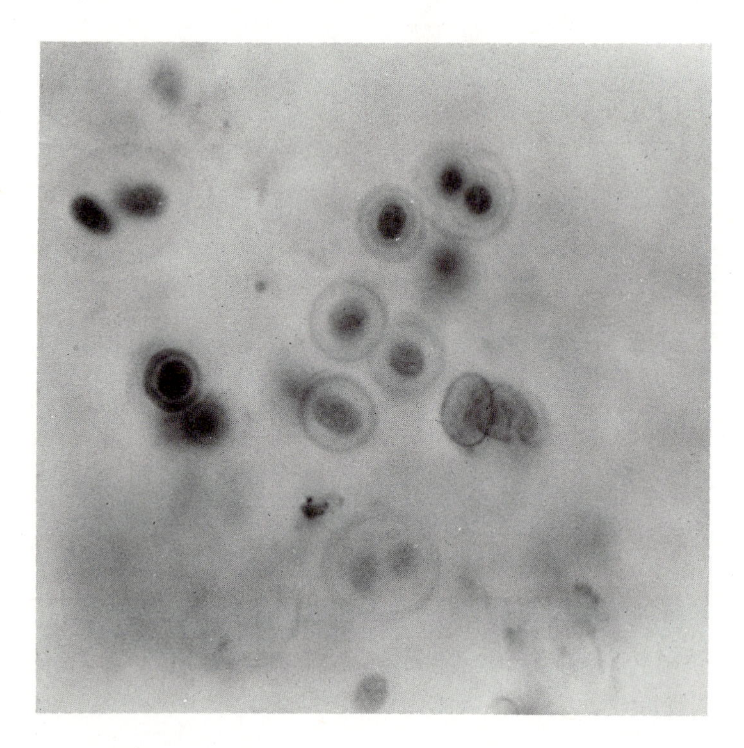

Fig. 2-1 Gloeocapsa, *a simple, unicellular, blue-green alga. Individual cells are surrounded by sheaths of a mucilaginous substance.* ×*550.*

small plant body consisting solely of a single, spherical cell.

Gloeocapsa, like the other blue-green algae, has an elementary internal structure. Instead of an organized nucleus with a limiting nuclear membrane, there is a dense mass of granular chromatin. No pigment-containing bodies (plastids) are found in the cytoplasm. The chlorophyll-containing structures are, instead, diffused throughout the periphery of the cellular contents.

In addition to a single kind of chlorophyll (chlorophyll *a*), *Gloeocapsa* contains yellowish pigments, a bluish pigment (phycocyanin), and a reddish pigment (phycoerythrin). The product of photosynthesis is not starch but glycogen, another carbohydrate.

Because of its lack of specialized internal organization and absence of motility, *Gloeocapsa* and other blue-green algae must be considered primitive. Their primitive structure does not imply, however, that the blue-green algae were ancestral to other groups of plants. In fact, some biologists believe that blue-green algae are not plants at all! These biologists point to the great similarity that exists between bacteria and blue-green algal cells. They feel that there is a closer relationship between these two groups of organisms—separated from the plant kingdom in some classification systems because they lack organized nuclei—than there is between the blue-green algae and other algal groups.

Motile Unicellular Algae

The unicellular blue-green algae are not the only one-celled algal organisms. Various other algal groups also contain organisms that consist of only a single cell. However, many other primitive unicellular algae are motile, using flagella as their principal means of locomotion. It is true that some of the green algae are unicellular and nonmotile, but it is presumed that in most of these instances their inability to move about is the result of a loss of flagella that these organisms possessed at some time during their evolution.

Among the flagellated unicellular algae, there seems to be a difference in kind and position of flagella that is correlated with differences in pigmentation and metabolism. Flagellar structure and position are often used to aid biologists in the classification of these primitive organisms.

Among the green plantlike organisms there is an entire division, the euglenophytes, that is characterized only by unicellular organisms. Plants such as *Euglena* (Fig. 2-2A), which zoologists may consider to be animals because they may exist without chloroplasts in certain environmental situations, are quite primitive in many respects, but they do show a degree of cellular specialization that is considerably more involved than that shown by the blue-green algae.

Not only do euglenophytes have an organized nucleus, but their photosynthetic pigments are contained within plastids that are limited externally by a membrane. A reddish eyespot, or stigma, is generally found at the anterior end. Some biologists suggest that the stigma is a light-receptive organelle that plays a part in the movement of the organisms in response to light, but there is no general agreement on this point. Also at the anterior end of the organism is a flagellum that propels the organism through the water. Although a second flagellum is present, it does not extend beyond the body of the organism.

The green algae, as well as the euglenophytes, include unicellular species that are propelled through the water by flagella. The number of flagella varies from species to species, but usually more than one is present. *Chlamydomonas* is unicellular. It is often considered to be one of the most primitive of the green algae and one that plays an important role in evolutionary studies of the green algae (Fig. 2-3). *Chlamydomonas* has two anterior flagella of the same length. At the periphery of the living material of the cell is a wall composed largely of cellulose. This rigid cell wall is a structural feature often found among plants and generally absent among species belonging to the animal kingdom. Within the cytoplasm is a single large chloroplast, often cup-shaped, and a single nucleus. These plants possess a reddish eyespot similar to that found in the euglenophytes.

The golden algae, another division of the algae, also has among its members some motile unicellular species. These organisms also have organized nuclei and plastids, and the cytoplasm is surrounded by a cell wall. In these species, however, two kinds of flagella are borne anteriorly. One is longer, extended forward, and has hairlike projections; the other is smaller and directed backward.

The motile unicellular algae discussed briefly on the preceding pages demonstrate the parallel development of similar features among algal groups that are not too closely related. Some members of all of these groups have flagellar locomotion,

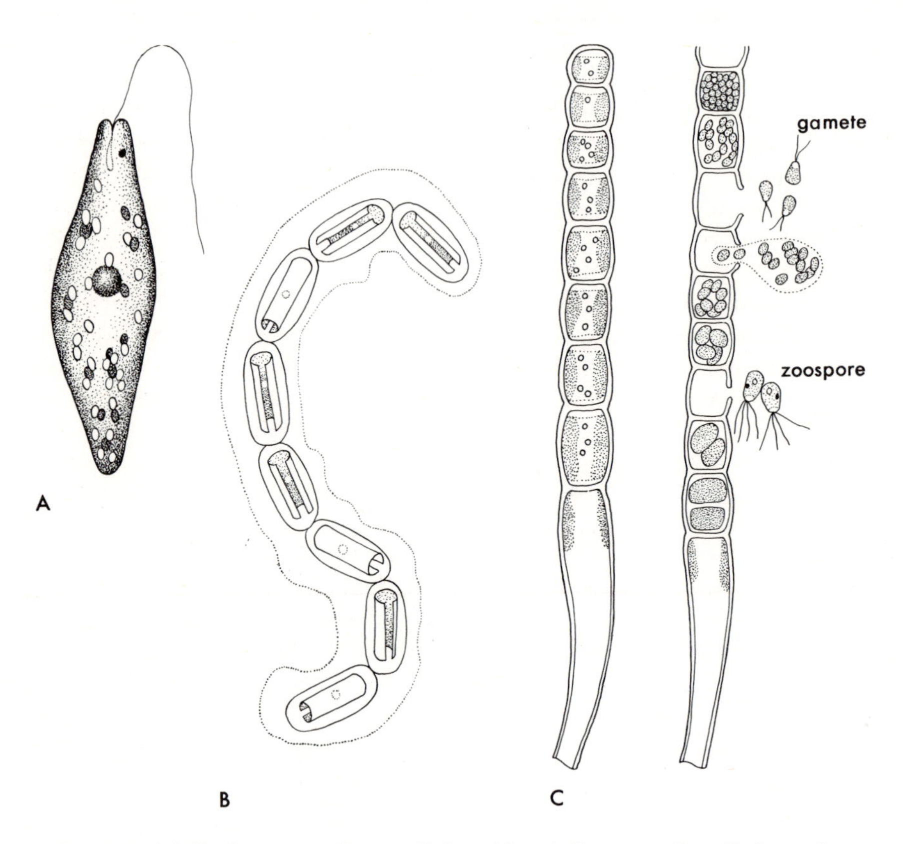

Fig. 2-2 (*A*) Euglena, *a motile, unicellular, chlorophyllous organism.* (*Redrawn from William H. Brown,* A Textbook of General Botany, *1925, published by Ginn and Company.*) (*B*) Geminella spiralis, *a filamentous colony of green algal cells.* (*Redrawn from G. M. Smith,* Fresh-Water Algae of the United States, *McGraw-Hill Book Company, 1950. By permission of the publishers*) (*C*) *Filaments of* Ulothrix, *a multicellular green alga.* (*Redrawn from L. H. Tiffany and M. E. Britton,* The Algae of Illinois, *University of Chicago Press, 1952. By permission of M. E. Britton.*)

and there is a parallel development of cytological features as well. This motile, unicellular form is well adapted to an aquatic environment and has been attained by various groups completely independently.

Flagella are not the only organelles used by motile unicellular algae to propel themselves. The golden algae include some one-celled species that have evolved a means of locomotion quite different from that of the flagellated forms. These species, called diatoms, are minute unicellular organisms that inhabit fresh and brackish water, as well as marine and moist terrestrial environments. Small samples of ordinary garden soil, if properly treated, will yield hundreds of these tiny plants whose distinctive feature is a cell wall composed of silica.

These "glass' shells are often intricately sculptured with fine dots or striations (Fig. 2-4). The enveloping silica shell is composed of two halves that overlap at the open faces, much like the two halves of a laboratory Petri dish. Many of the diatoms have various slits or pores in the cell wall. A crystalline material that becomes fibrillar as it takes up water is secreted through these openings. The fibrils contract when they come in contact with a solid substrate, thereby pulling the unicellular organism along the surface.

In all of these unicellular organisms it is obvious that motility in itself is no indication of relationships among the forms involved. Development of a means of locomotion is more likely an adaptive feature with considerable survival value in an aquatic environment. Furthermore, the ability of certain algae to propel themselves through the water was attained in more than one way, in more than one group, and, most likely, at more than one time in the history of the plant kingdom.

CELL WALLS

Most of the unicellular photosynthesizing organisms have a more or less rigid cell wall at the periphery of the protoplast. Some species—for example, members of the euglenophytes—have a much less rigid membrane, which is not, strictly speaking, a cell wall. Most likely the wall around cells is a derived feature, the most primitive organisms having lacked it. In fact, one of the evolutionary steps that occurred early in the separation of the animal and plant kingdoms was the formation of walls around cells in the plant line, and the failure of such a development among animal organisms. This characteristic, although a relatively minor one in the evolution of unicellular organisms, is of major significance when the evolutionary directions of the two kingdoms are considered. Because animals lack cell walls, an animal that attained a certain size and that was not supported by a buoyant, aqueous environ-

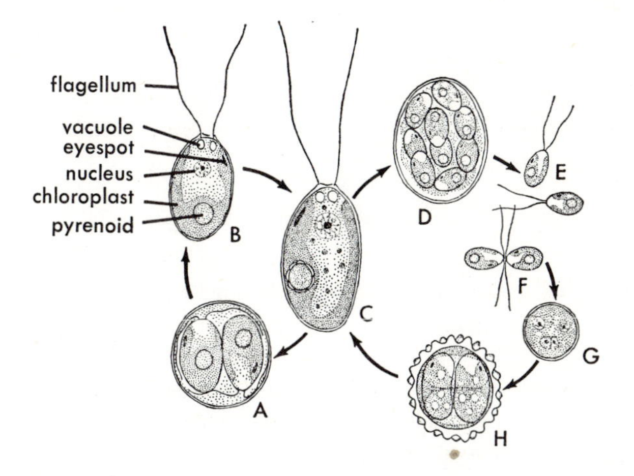

Fig. 2-3 Chlamydomonas, *structure and life cycle. (From* The Plant World, Fourth Edition, *by Harry J. Fuller and Zane B. Carothers. Copyright 1941, 1951, © 1955, 1963 by Holt, Rinehart and Winston. Reprinted by permission of Holt, Rinehart and Winston.)*

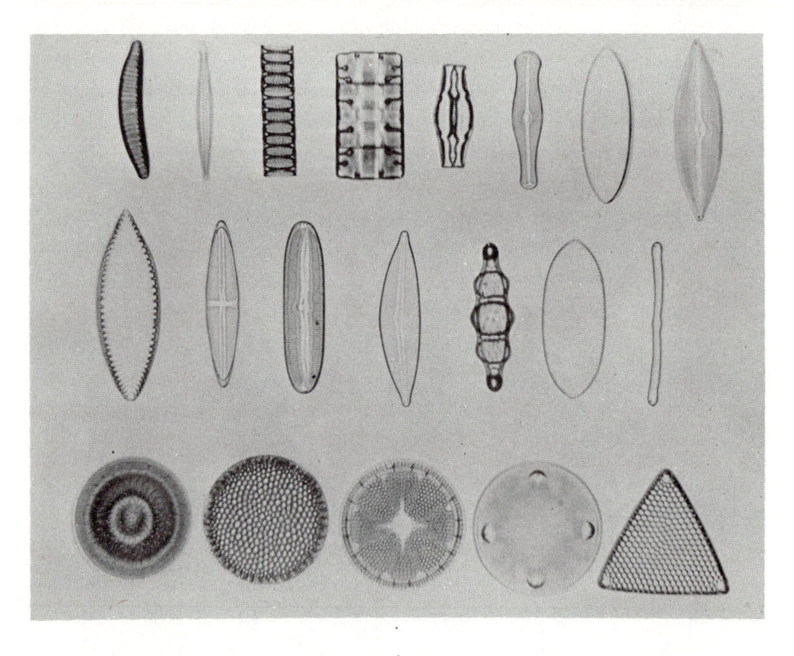

Fig. 2-4 *Siliceous shells of an assortment of diatoms. ×160.*

ment could not exceed this size without an internal skeletal system. Vertebrates, for example, may reach tremendous sizes because support is effected by a skeleton. Plants, on the other hand, have no structure quite comparable to internal skeletons, and rigidity is the result of the aggregate strengths of individual cell walls. Redwood trees exceed 350 feet in height, and only cell walls, particularly those of the wood cells, are responsible for their ability to stand erect and to survive winds and storms.

Primitive cell walls were undoubtedly composed of simple carbohydrates—primarily cellulose and pectic substances. Later, cell walls became more complex, with additional substances, such as lignin, suberin, silica, and other compounds being incorporated.

CELLULAR SPECIALIZATION

Although quite different from complex plant forms in structure and function, unicellular organisms must have been the ancestors of these higher plants. Each unicell, of course, is an entire plant completely equipped for all of its functions. Within a single protoplast lie the facilities for respiration, photosynthesis, reproduction, motility, digestion, and a host of other processes. As organisms became more complex, many of these functions were confined to certain cells or groups of cells, while other cells carried on different functions. In a tree, for example, only some of the cells are involved in photosynthesis. Water transport is carried on mainly in specialized conducting cells that are actually dead. Reproduction occurs only in localized regions.

Unicellular organisms, therefore, probably represent the most complex type of single cell. From these elaborate cells were derived the higher organisms, more complex in overall structure, but with individual cells more specialized (less generalized) than those of unicellular organisms. Individual cells in multicellular organisms are each genetically like all others in the same organisms.

ORIGIN AND EVOLUTION OF REPRODUCTIVE SYSTEMS

The ability to reproduce itself is one way in which a living organism differs from a nonliving system. The simplest, most primitive, unicellular plants have the capacity to produce more individuals identical with the parent. But even within unicellular plants we can see an evolutionary sequence leading to progressively more complex reproductive systems.

Members of the blue-green algae exhibit an extremely simple means of reproduction. A *Gloeocapsa* cell, for example, produces more cells like itself simply by dividing. This organism contains no organized nucleus, and recognizable chromosomes are not evident, but in all likelihood the chromatin material in the cell divides equally at the time of cell division. Nucleic acid replication probably occurs before cell division, as it does in plant cells with more highly organized nuclei. After a *Gloeocapsa* cell has divided to form two cells, the daughter cells are held loosely together in a mucilaginous sheath. No cytoplasmic connection exists between the daughter cells, so we may still regard the organism as unicellular. So far as we know, there is no evidence of sexual reproduction among the blue-green algae.

The euglenophytes also reproduce by a simple fission of cells. In these unicellular species, the nucleus divides mitotically, and two similar daughter individuals are the result of cell division. There is no convincing evidence of sexual reproduction among the euglenophytes either.

Members of the green algae may be used as examples of sexually reproducing unicellular organisms. The unicellular *Chlamydomonas* has both sexual and asexual means of reproduction (Fig. 2-3). Asexual reproduction involves the division of a cell into two, four, or eight daughter cells. Each of these daughter cells subsequently develops paired anterior flagella, is released, and swims away as a new individual. In some instances the daughter cells do not develop flagella, and the products of the division of the original cell remain embedded in the gelatinized matrix of the cell wall of the parent. Further divisions of the embedded daughter cells may occur, producing a loosely aggregated colony of individual organisms. Because this aggregation of *Chlamydomonas* cells may resemble the typical configuration of the green alga *Palmella*, the nonmotile, colonial stage of *Chlamydomonas* is frequently called the *palmelloid stage*. When conditions are more favorable, the daughter cells in the aggregation develop flagella, and the individual organisms then swim away.

***Sexual Reproduction in* Chlamydomonas** Sexual reproduction in *Chlamydomonas* involves the pairing of individuals that, in effect, function as *gametes.* When protoplasts are released from the cell walls, two gametes fuse at their flagellar ends to form a diploid fusion cell, the *zygote.* Under favorable conditions the contents of the zygote divide meiotically, producing four motile cells, each with half the number of chromosomes of the zygote or the same number as in the parent plant and in gametes. In this way the usual vegetative chromosome number is restored.

Most species of *Chlamydomonas* have gametes of only one structural type; fusion of these similar cells is called *isogamy.* Some species have gametes of two sizes and are said to have a *heterogamous* sexual process in which smaller, motile gametes fuse with larger, motile gametes. It must be emphasized that the isogamous individuals referred to here are designated isogamous on the basis of morphological similarity alone. In other words, when they are simply observed, the two gametes appear similar. These gametes may be functionally heterogamous, however.

When heterogamous gametes are involved, sexual fusion will occur only if gametes from two *mating types* come in proximity. Naturally, it is impossible to refer to one of these gametes as "male" and the other as "female"; instead, it is common to call one arbitrarily the "plus" (+) strain and the other the "minus" (−) strain.

An extreme development of heterogamy occurs in at least one species of *Chlamydomonas.* In this species one entire individual functions as a female gamete and is joined by a smaller, motile cell produced by divisions of the protoplast of another individual. The larger cell, which loses its motility, is the *egg.* The motile reproductive cell may be called a *sperm.* This type of reproductive system in which a sperm cell fertilizes an egg cell is called *oogamy.*

Within one genus, therefore, it is possible to see sexual processes ranging from one thought to be the most primitive (isogamy), to a more advanced type (heterogamy, with both gametes motile), to extreme heterogamy (oogamy).

Chlamydomonas is of further interest to biologists because this genus demonstrates the possible origin of sexual reproduction, at least among some of the algae. In certain instances, after the protoplast of a cell has divided, the new cells are released and may be regarded as new individuals. In other situations the products of these cell divisions may fuse with each other and act as gametes. No difference has been observed between motile daughter cells that function as new individuals and those that function as gametes. It appears therefore that sexual reproduction need not occur, and the daughter cells may be regarded as asexually produced progeny.

Gametes and asexually produced offspring are homologous (that is, morphological equivalents) only in cases of isogamy. In the heterogamous species the daughter cells that swim away as new individuals are all of one type, whereas those cells destined to become gametes are of two types. This difference is even more pronounced among the oogamous forms.

Although *Chlamydomonas* may be regarded as a primitive green alga, its modes of reproduction are basically identical to those in a large number of algae.

Chlamydomonas cells are always in the haploid phase with one exception, the zygote, which is produced after fusion of gametes. (The first division of the zygote is meiotic, however, and the haploid condition is restored). Other, more highly organized algae reproduce in the same manner, although in many-celled algae, gametes are produced exclusively in certain specialized cells, called *gametangia* (singular, *gametangium*). But even among these algae, gametes are generally produced within one cell by a division of the protoplast and, in isogamous and certain heterogamous forms, may be motile. In oogamous algae, the entire contents of the gametangium may function as a nonmotile egg; in some instances, more than one egg is produced after division of the protoplast.

Among the more complex algae, however, gametes function only as gametes and are distinct from motile, asexually produced spores that germinate to form new individuals. There is, then, a specialization of function of certain of the cells.

A generalized life cycle of an alga such as *Chlamydomonas* (Fig. 2-3) or of other algae with a life cycle practically entirely in the haploid phase involves a haploid individual that may produce motile, haploid, asexual spores that germinate directly into new individuals. These motile spores are called *zoospores.* The same haploid plant may also produce haploid gametes that fuse in pairs to form a diploid zygote. Either immediately or after a resting period of some length the zygote undergoes a meiotic division that again restores the haploid chromosome condition, and four haploid spores (generally zoospores) are formed. Each of these produces a new plant. In *Chlamydomonas* and other genera the new individual consists of only one cell, but new plants are composed of many cells in more advanced kinds of algae.

Although in many algae a haploid individual may produce both gametes and zoospores, in a number of them production of these two types of cells occurs in two different types of individuals. One type of plant of a certain alga may produce only gametes in its lifetime; others produce only zoospores. Often the production of gametes or zoospores is determined by environmental factors, or by the age of the plants.

Among the higher green plants that live on land, however, spores are produced exclusively by the diploid phase of the plant and gametes are produced by another phase of the plant's life cycle, the haploid phase. Evolution of the sexual process apparently has advanced to such a stage among the green plants other than algae that there is considerable distinction between spore and gamete, even though these structures might be considered to be homologous, that is, identical at an earlier stage of their evolutionary history. Furthermore, there is also considerable distinction between the two phases of the life cycle in the plant that produces these two types of reproductive cells. In typical situations in these higher plants a spore cannot function as a gamete, nor can a gamete produce a new individual directly.

Other advances are found in the reproductive processes in the higher green plants. Among the algae, gamete-producing structures are typically single cells. In the bryophytes and vascular plants, there are more elaborate gametangia, all multicellular. Oogamy is the only type of sexual reproduction found in the bryophytes and vascular plants. This topic will be further elaborated in the discussion of the land plants in Chapter 4.

ORIGIN OF MULTICELLULARITY

Obviously, more than sexual specialization is involved in the evolution of complex higher plants from presumed ancestral types such as *Chlamydomonas.* A *Chlamydomonas* plant consists of only one cell, and in that cell all of the functions of the organism are carried out. In an advanced type of plant (for example, a pine tree) there are numerous cells of many kinds, and there is a difference among cells both in appearance and in function. To derive a complex vascular plant from the algae, not only must the number of cells be increased, but different functions must become relegated to different cells. The development of such a condition of multicellularity is not understood completely, but there are trends found among the green algae that suggest certain of the steps that may have been involved.

We shall now describe one hypothesis about the origin of *multicellularity.* The *palmelloid* stage of *Chlamydomonas* has been mentioned already. When a single *Chlamydomonas* cell divides and the daughter cells produced do not form flagella but remain loosely aggregated in the gelatinized wall of the parent individual, a loose aggregate is formed. These daughter cells also divide, and thus numerous nonmotile cells eventually become associated within a gelatinous mass. Under some conditions the daughter cells may develop flagella and swim away, breaking up the aggregate.

Cells of the genus *Palmella* are nonmotile during most of their existence. Repeated divisions of a single cell result in a loosely associated colony of cells. At certain times some of the cells may develop flagella, swim away, and settle down. After repeated divisions of these cells a new colony is formed. There seems to be no question about a relationship between *Palmella* and *Chlamydomonas,* since the former is the result of an evolutionary trend toward loss of motility and aggregation of unicellular organisms into colonies.

Filamentous Structure in Algae

An advance in this evolutionary trend toward multicellularity would involve aggregation of these nonmotile unicellular organisms into some kind of regular structure—a linear filament, for example. Just such a condition exists in the green alga *Geminella,* which consists of cells loosely lined up within a cylindrical mucilaginous sheath (Fig. 2-2B). New colonies are formed simply by the breaking apart of the loose aggregation of cells within one cylinder.

As the next step in the development of this tendency toward multicellularity, we might expect that aggregation would involve a more closely knit filament, with cells actually attached to adjacent ones. In fact, this type of structure, the filamentous construction, is the most common type of algal body. A large percentage of the green algae are filamentous. Some species are branched, others unbranched.

Ulothrix (Fig. 2-2C), a common example, is an unbranched filament of haploid cells. One cell at the end of the filament, the "holdfast cell," is modified into a structure that attaches the rest of the filament to the substrate. Contents of certain of the cells may divide a number of times to produce up to 32 motile cells, each of which resembles a *Chlamydomonas* cell. Each motile zoospore cell of *Ulothrix* has four flagella, however. These zoospores produce new filaments directly. Other cells of the *Ulothrix* filament may divide to produce as many as 64 biflagellate gametes. These too resemble *Chlamydomonas* individuals quite closely. All *Ulothrix* gametes are alike morphologically, but apparently there is some physiological difference between gametes from different plants, for gametes from the same plant will not unite with each other. The quadriflagellate diploid zygote formed after the union of gametes may be motile for a time, but after a resting period it undergoes meiosis and produces as many as 16 motile or nonmotile haploid spores that germinate to form new plants.

Although a plant such as *Ulothrix*, mentioned here simply as a typical green algal type, is quite distinct from a simple plant such as *Chlamydomonas* and much more complex in body structure, it is still possible to recognize affinities between the two genera. For example, the motile cells—zoospores and gametes—of *Ulothrix* are almost indistinguishable from a *Chlamydomonas* cell. The more complex body structure of *Ulothrix* may have evolved from a *Chlamydomonas*-type individual in a series of stages involving first a loose aggregation of the nonmotile phase, then a more regular but still loose alignment of cells into a filamentous form, and, finally, an even closer alignment of cells with adjacent cells actually connected by cytoplasmic strands within a filament. Thus in *Ulothrix* the adjacent cells are more than just physically in contact. *Ulothrix* is not simply a linear colony of cells but a filamentous *whole*. Although we cannot prove that these events actually occurred in the evolution of multicellularity, the sequence is a logical one and is supported by a number of examples among the algae.

The differences in structure between *Ulothrix* and *Chlamydomonas* serve to point out an evolutionary principle seen again and again in the plant and animal kingdoms. Although *Ulothrix* is more complex in structure, sexual reproduction in *Ulothrix* is isogamous, the reproductive mode considered to be the most primitive. In contrast, at least one species of *Chlamydomonas*, while retaining a very simple type of body form, is oogamous, a more advanced type of reproduction. In evolution, change of various parts or processes in individuals do not proceed at the same rate. In most instances, different types of changes are independent of each other. An organism may be relatively advanced in one respect while retaining certain features that have not evolved beyond a primitive stage. Although it is suggested that *Chlamydomonas* could represent an ancestral type of structure from which more advanced types of algae evolved, the actual ancestor probably was not *Chlamydomonas* itself but some other unicellular form that resembled *Chlamydomonas* in more ways than it resembled *Ulothrix*.

A further elaboration of the filamentous habit in algae is the development of branches (Fig. 2-5D). Branching is found in many green algae and is of some significance in attempts to determine the origin of more highly specialized land plants from algal ancestors. The most extreme elaboration of the filamentous body type, a branched filament with a thickness of more than one cell, is found occasionally, but not frequently, among the green algae (Fig. 2-5E).

If the evolutionary sequence leading to a filamentous type of construction is carried still further, we might postulate that a platelike algal body might be formed

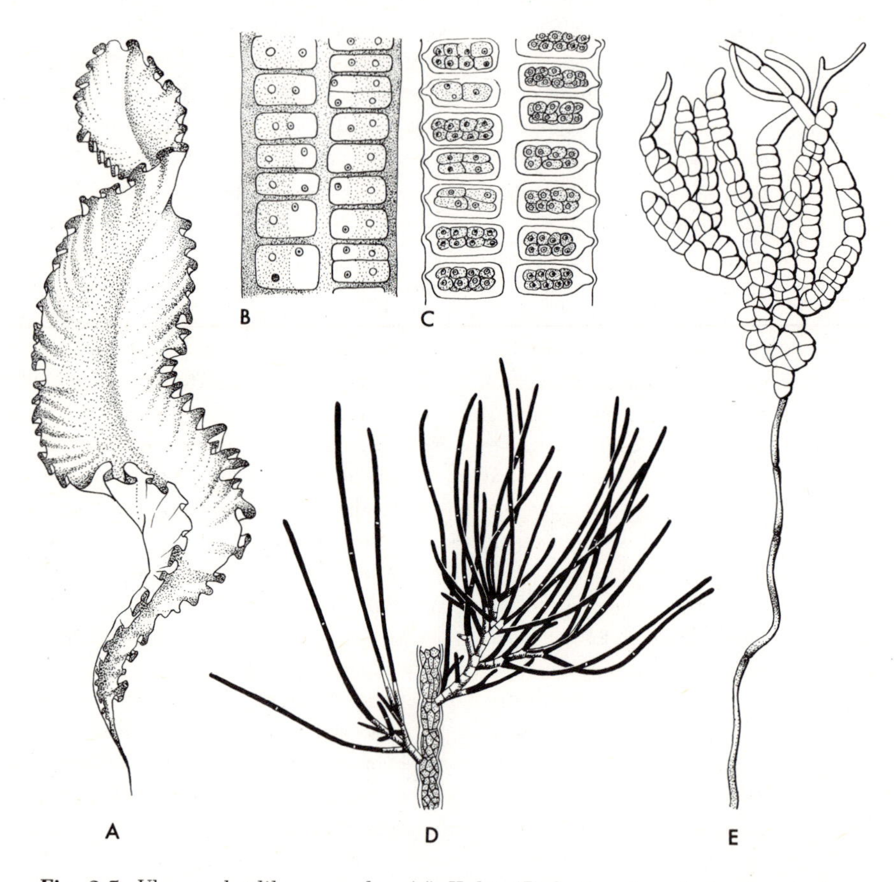

Fig. 2-5 Ulva, *a platelike green alga. (A) Habit. (B) Section through the plant body. (C) Division of cellular contents in the production of reproductive cells. (Redrawn from G. M. Smith,* Cryptogamic Botany, *vol. 1, McGraw-Hill Book Company, 1955. By permission of the publishers.) (D) Fragment of a plant of* Draparnaldia, *a branched green alga. [Redrawn from* The Plant Kingdom, *by William H. Brown, © Copyright, 1935, by William H. Brown. Used by permission of the publisher, Ginn and Company (Xerox Corporation.)] (E)* Fritschiella, *a green alga with a branching habit and three-dimensional fleshy body form. (Redrawn from M. O. P. Iyengar,* New Phytologist, *vol. 31, p. 331, 1932.)*

by cell divisions in two planes. Examples of that kind of structure are also abundant among the algae. A very common form is the marine sea lettuce, *Ulva,* which is composed of a plate of photosynthetic cells only two cells thick. The sea lettuce is of interest because of its characteristic life cycle. Some of the cells in a haploid plant divide to produce haploid, motile gametes. These fuse in pairs to produce diploid zygotes. In *Ulva* (Fig. 2-5A–C), reduction division does not occur immediately upon germination of the zygote; instead, a series of mitotic divisions produce a new plant that looks identical with the haploid phase but is diploid throughout. Reduction division occurs in some of the cells of the diploid phase, producing motile haploid zoospores that germinate into new haploid individuals. In *Ulva* as well as in many other algae the multicellular haploid phase alternates with a multicellular diploid phase. This alternation and its implications will be discussed again in relation to land vascular plants in Chapter 4.

Not all plants with alternating haploid and diploid stages have two identically appearing phases. In most cases the gamete-bearing stage differs in structure from the spore-bearing stage.

A MOTILE MULTICELLULAR EVOLUTIONARY SERIES

Another evolutionary series tending toward a multicellular body type is found in the motile green algae. This trend apparently had limitations in evolutionary potential in the plant kingdom. *Gonium* (Fig. 2-6A, B) a green algal genus that is found as colonies of 4 to 16 *Chlamydomonas*-like cells arranged in a slightly curved or almost flat plate. All cells in this colony swim in unison, so the entire plate moves as a unit. Other genera are more advanced in form, having 16 or more motile cells arranged in an ellipsoidal colony with flagella extending out from the periphery. The ultimate development in this evolutionary series of motile aggregates of cells is *Volvox* (Fig. 2-7). *Volvox* organisms form a spherical mass that sometimes has thousands of cells arranged at the periphery. These spheres may be large enough to be seen with the naked eye. In some species of *Volvox* there is actually cytoplasmic connection between adjacent cells, and thus it is more appropriate to refer to these spheres as multicellular individuals than as colonies. Not all cells of a *Volvox* have the same function; some cells are purely vegetative, some produce motile male gametes, some produce eggs, and still others produce new spheres vegetatively.

Although this type of multicellular structure may be advantageous in certain environments, it has limitations with respect to potential size and suitability as a possible precursor to more highly advanced land plants. The evolutionary sequence, often referred to as the "volvocine tendency" (after the genus *Volvox*), is of considerable interest, however, in demonstrating another path followed by green algae through which multicellular individuals may have evolved from motile unicellular organisms.

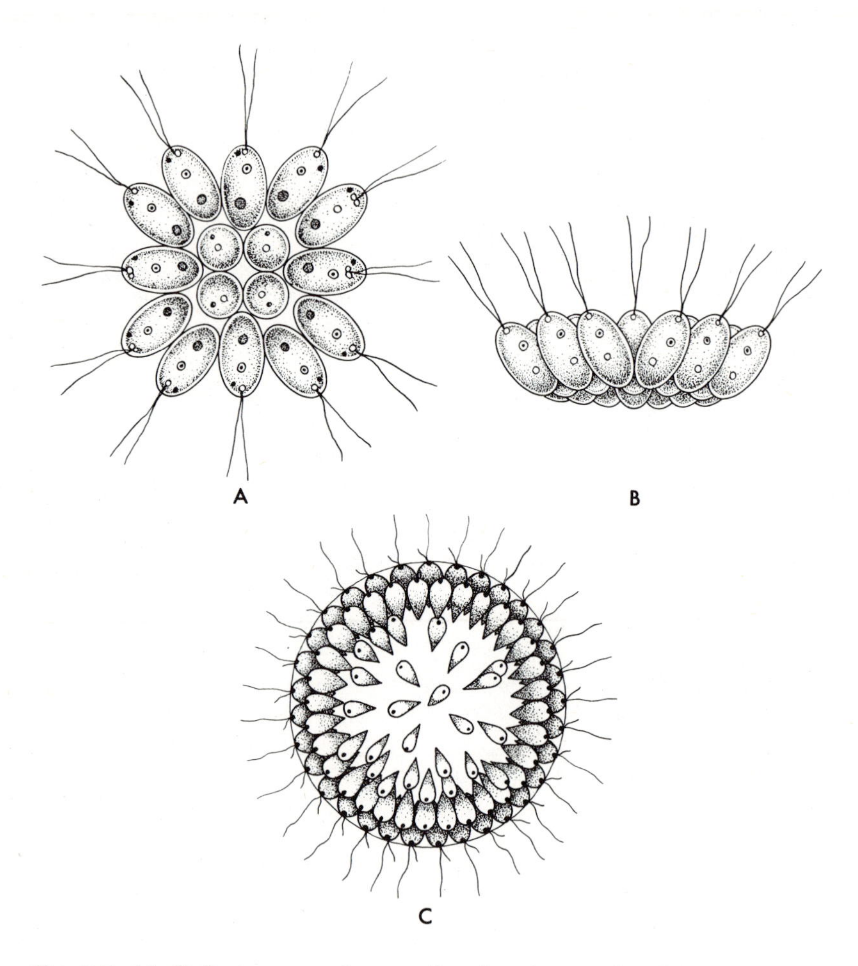

Fig. 2-6 (*A, B*) Gonium pectorale, *a motile, colonial green alga.* [*Redrawn from* The Plant Kingdom, *by William H. Brown,* © *Copyright, 1935, by William H. Brown. Used by permission of the publisher, Ginn and Company* (*Xerox Corporation.*)] (*C*) Uroglena, *a motile, colonial golden alga.*

The Tubular, Nonseptate Trend Another evolutionary trend seen among the green algae and paralleled in other algae and some fungi is the development of a filamentous plant body with few, if any, partitions. A single cell elongates, forming a tubular, possibly branching structure, with nuclear divisions taking place throughout this process. At maturity a very large number of nuclei may be found in the filament. No cytoplasmic divisions occur, however, and cross walls are generally missing, except at a few places in the filament where reproductive structures are formed. Rather elaborate algal bodies frequently are built on this tubular theme (Fig. 2-8), but this body type is carried to even further extremes in certain fungal groups.

SUMMARY OF SOME EVOLUTIONARY TRENDS IN THE GREEN ALGAE

The preceding short survey of evolutionary trends in the green algae is an attempt to demonstrate how progressive specialization occurs in body type, in mode of reproduction, and in relative development of the haploid and diploid cytological phases. The green algae represent a group of organisms that demonstrate many combinations of features. There is no single line of evolution involving elaboration of body structure that coincides with the development of more advanced sexual reproduction, and so forth. Simple unicellular forms may be oogamous, whereas complex branching forms may be isogamous. Alternation of multicellular phases may be present or absent in forms with isogamy; similarly, oogamous forms may or may not have regularly alternating haploid and diploid multicellular phases. Nevertheless, the individual evolutionary trends are discernible, and the fact that in a given alga not all characters are in the same stage of evolution serves to accentuate the principle that in organisms evolution of all parts need not, and does not, progress at the same rate.

PARALLEL TRENDS IN OTHER ALGAL GROUPS

In all groups of algae many of the evolutionary trends, both in vegetative and reproductive aspects, seen in the green algae are duplicated. Some of these trends are somewhat less developed in other groups, but others may be more pronounced. Distinguishing differences among the major groups of algae are not primarily morphological, but biochemical, concerning, for example, types of pigments and types of metabolic processes. Structure and placement of cilia or flagella are also significant.

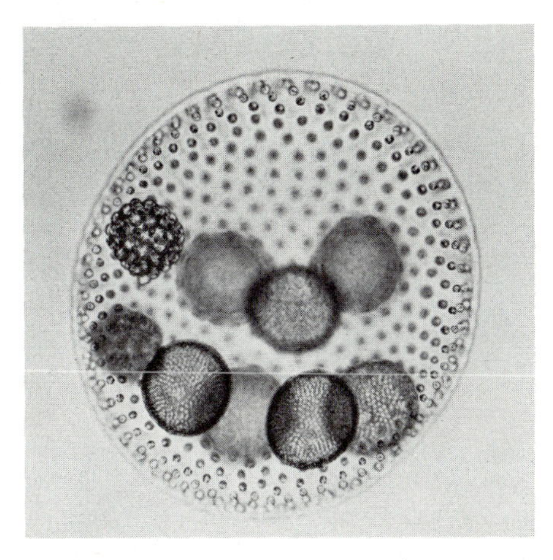

Fig. 2-7 Volvox, *a large, globose, motile green alga composed of numerous cells. Note seven new spheres produced asexually and two male spheres.* ×94. (*Photograph courtesy of Richard C. Starr.*)

Fig. 2-8 Caulerpa, *a large, unicellular, multinucleate green alga.* ×*1.5. (Photograph courtesy of Michael J. Wynne.)*

The golden algae, for example, contain motile unicellular species that closely parallel *Chlamydomonas* and its relatives among the green algae. Although types of flagella and their placement differ in these two groups, the differences in kinds and proportions of plastid pigments are more important (note the distinguishing characteristics in the classification presented in Chapter 1).

The golden algae contain species composed of motile cells linked together similar to those in the volvocine line in the green algae (Fig. 2-6C). Furthermore, the filamentous mode of body construction is evident, apparently having been derived in the same manner as in the green algae. Tubular, nonseptate golden algae are also known. As in the green algae, reproduction in the golden algae ranges from isogamy to oogamy.

There are no unicellular forms among the brown algae. All are constructed on the filamentous pattern, although many species get to be quite massive. In fact, some of the brown algae growing in the oceans reach lengths of a few hundred feet. Internal organization of the brown algae is the most elaborate and highly specialized of all the algal groups. The main axes of some brown algae contain elongated cells closely paralleling phloem cells, the food-conducting cells of vascular plants. In structure, certain brown algae superficially resemble higher vascular plants, and even live near the shore where they are periodically exposed to dry air. In their combination of features, the brown algae would seem to be the most likely precursors to land plants: the form of the plant body is right, there is the proper kind of sexual reproduction, and the alternation of cytological phases in some brown algae is very similar to that of land plants.

The chemistry of the brown algae, however, gives rise to the greatest objection to considering them as land plant precursors. Absence in the brown algae of chlorophyll *b* and the occurrence of chlorophyll *c* and a large percentage of xanthophylls distinguish them from the green algae and higher green plants. Nevertheless, the brown algae are interesting as another example of parallel development in more than one group of unrelated organisms.

Evolution of methods of sexual reproduction among the brown algae follows the familiar pattern of change from isogamy to heterogamy, with many species being oogamous. Most brown algal genera have an alternation of haploid and diploid multicellular phases (Fig. 2-9). These two phases may be alike (as in *Ulva* of the green algae), or they may be quite different.

Finally, many parallels to the brown algae are found among the red algae. Only two genera of unicellular forms are found among the red algae. In both the red and brown algae, the typical construction of the plant body is a modification of the filamentous habit. These species may be radially symmetrical and often elaborately branched, or they may be platelike and flattened, with bilateral symmetry. No flagellated reproductive cells are known, and all forms reproduce sexually by oogamy.

Among the red algae the alternation of phases is, at times, complex. Instead of a precise alternation between a haploid and a diploid phase, an additional diploid phase may be intercalated. As an example, the genus *Nemalion* may be cited (Fig. 2-10). *Nemalion* is a slender, filamentous organism with some branches specialized for production of male gametes and others for the production of female reproductive organs. After the male gamete comes in contact with the female reproductive organ, a number of short branches that produce nonmotile spores in their terminal cells arise at the base of this organ. It is assumed that these spores are diploid and that the filaments that produce them are results of equational divisions of the diploid zygote. From these diploid spores arise small, delicate filaments. Certain cells of these filaments divide, possibly meiotically, to produce four spores in each. Although all details of the life cycle of *Nemalion* are yet to be worked out, it is assumed that these spores germinate to produce a new haploid phase of the plant. This life cycle is summarized in Fig. 2-11A.

Polysiphonia is another genus of red algae that has a life cycle essentially similar to *Nemalion*. In this genus, however, meiosis is delayed beyond the point where it is assumed to occur in *Nemalion*. Diploid spores, produced by short, diploid filaments that arise after fertilization are released and germinate to produce new diploid plants. These plants are similar to the male and female haploid plants and occupy a considerably more significant part of the life cycle than do the small diploid filaments of *Nemalion*. Meiosis occurs in certain of the cells of the diploid individual, forming haploid spores, each of which germinates to form a new haploid individual. The type of alternation found in *Polysiphonia* and certain other red algae is not

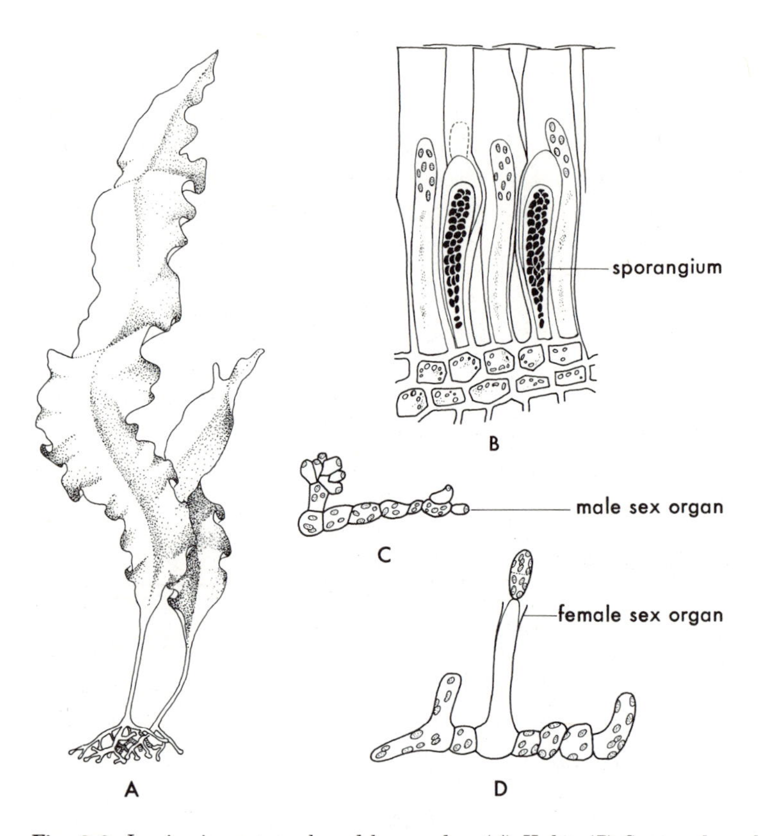

Fig. 2-9 Laminaria, *a strap-shaped brown alga. (A) Habit. (B) Section through part of the diploid phase showing elongated sporangia with spores within. (C) Haploid phase producing male gametes. (D) Haploid phase producing female gametes. [(A) and (D) redrawn from* General Botany for Colleges *by Ray E. Torrey, Copyright 1922, 1925, 1932 by R. E. Torrey. Reproduced by permission of Prentice-Hall, Inc. (B) redrawn from F. Oltmanns,* Morphologie und Biologie der Algen, *vol. 2, Gustav Fischer, 1922. (C) redrawn from various sources.]*

common in the plant kingdom and is of less evolutionary significance than the regular alternation of a haploid and a diploid phase that typically occurs in green land plants.

EVOLUTIONARY POTENTIAL OF THE ALGAE

From the preceding discussion of the algae, it should be clear that the group is heterogeneous in many ways. There are many morphological types, reproductive mechanisms, alternating life-cycle phases, pigment types, and so forth. Rather than follow the older systems of classification in which the "Algae" were listed as a phylum or division, it is probably more exact to consider the term "algae" only as a convenient means of collectively referring to these chlorophyll-containing, nonvascular plants. Within this assemblage are several distinct lines, each delimited rather nicely by certain features of pigmentation and metabolic storage products and, to some extent, by characteristics of flagellation. The various algal tendencies developed to varying degrees among the major algal groups are apparently examples of parallel or converging evolution.

Most of the algae attained characteristics that were not adapted to further evolutionary advance, especially in connection with a new type of environment on land. A few, however, succeeded in making the transition from an aquatic habitat to a terrestrial one. It will be seen in Chapter 4 that among the diverse structural and biological adaptations found among the algae, those that we feel must have been

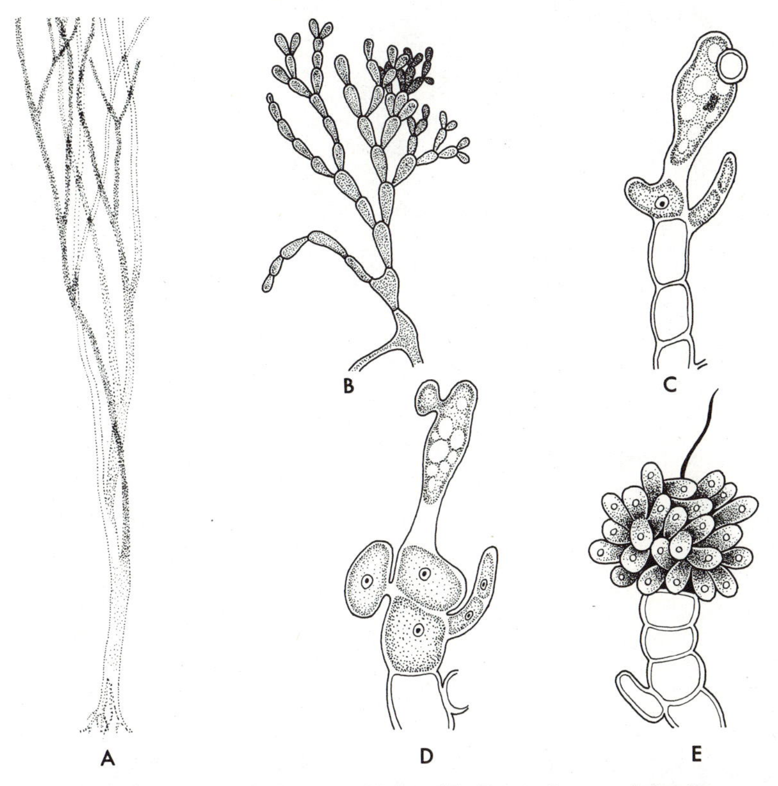

Fig. 2-10 Nemalion, *a red alga.* (*A*) *Habit.* (*B*) *Terminal part of the filamentous branching portion with male sex organs at the tips.* (*C*) *and* (*D*) Batrachospermum, *a red algal genus with a sexual process similar to that of* Nemalion. (*C*) *Male gamete* (*sphere near tip*) *in contact with the female sex organ.* (*D*) *Fusion of contents of male gamete with female sex organ* (*near tip*). *Beginning of outgrowth from the base of the female sex organ is at the left.* (*E*) Nemalion, *production of spore-bearing filaments from the base of the female sex organ.* [(*A*), (*B*), *and* (*E*) *redrawn from* General Botany for Colleges *by Ray E. Torrey, Copyright 1922, 1925, 1932 by R. E. Torrey. Reproduced by permission of Prentice-Hall, Inc.* (*C*) *and* (*D*) *redrawn from H. Kylin,* Berichte der Deutschen Botanische Gesellschaft, *vol. 35, p. 160, 1917.*)]

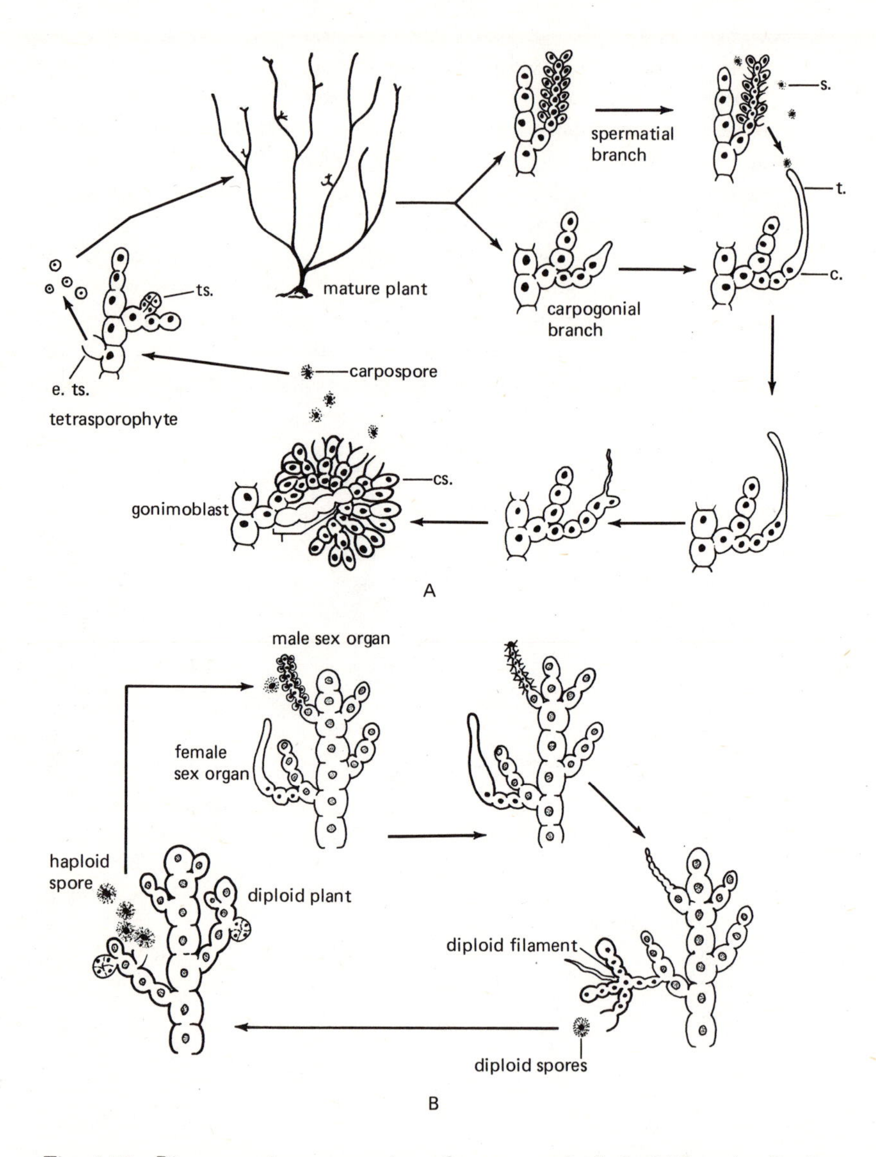

Fig. 2-11 *Diagrammatic representations of two types of red algal life cycles. Portions with fine outlines are thought to be in the haploid chromosome condition; portions with heavier outlines are presumed to be diploid.* (*A*) Nemalion *type of life cycle.* (*B*) *Life cycle characterized by* Polysiphonia. (*Redrawn from* Introductory Botany, *by Arthur Cronquist. Copyright © 1961, 1971 by Arthur Cronquist. By permission of Harper & Row, Publishers.*)

prerequisites for ancestors of green vascular plants are found among the green algae. Most readers are familiar with the green film on tree trunks and moist stones in a forest. This green scum is actually composed of countless minute, unicellular green algae that are completely suited to a life on land. Certain other algae are successful outside the aquatic habitat as parasites on the leaves of higher vascular plants. Furthermore, in the soil are numerous algal forms that thrive with only a small amount of water and can persist in a resting condition through adverse desiccating conditions.

These so-called "terrestrial" algae, however, have nothing to do with the species that gave rise to land vascular plants. In most cases the algal forms living out of water are simple and often unicellular. In some instances it is suspected that the simplicity, rather than being an inherently primitive condition, is a result of a reduction of more complex characteristics.

Though many algae represent "dead ends" as far as subsequent evolutionary development is concerned, there is no doubt that they are extremely successful forms, since they are able to exist in a wide range of habitats all over the earth. Furthermore, the great diversity among the algae would suggest that many forms among these plants would be able to adapt themselves to changing environmental conditions.

ALGAE AS EXPERIMENTAL ORGANISMS

A number of botanists utilize algae in various experimental disciplines because of the suitability of these organisms for elucidating certain fundamental biological principles.

Some unicellular organisms, notably the nonmotile genus of green algae, *Chlorella*, are important in the study of photosynthetic processes. These algae are easily cultured. Because of their simple structure, it is relatively easy to extract pigments or intermediate products of photosynthesis from them.

Interesting experiments in mechanisms of heredity, differentiation, and cell physiology have been carried out on the green alga *Acetabularia*, a genus of relatively large size (several centimeters tall) which has only a single nucleus in its vegetative phase. By grafting part of one plant to part of another plant of a different *Acetabularia* species, it is possible to determine the relative roles of nucleus versus cytoplasm in inheritance or in controlling various cellular processes.

Developmental studies are being carried out on the fertilized egg of the common brown alga *Fucus*. The process of fertilization can be observed in this genus, and an attempt to follow the events immediately following fertilization is being made.

Other algae also serve as excellent experimental organisms. The principal advantage of these plants for experimental work is their general lack of structural complexity. Investigators thus can concentrate on the structure or function in which they are interested without the additional complications found in more highly specialized plants.

FURTHER READING

Bold, H. C., *Morphology of Plants.* 3d ed. New York: Harper & Row, 1973.

Bold, H. C., *The Plant Kingdom,* 4th ed. Englewood Cliffs, N.J.: Prentice-Hall, 1976.

Chapman, V. J., *The Algae.* London: Macmillan, 1962.

Chapman, V. J., *Seaweeds and Their Uses.* London: Methuen, 1970.

Doyle, W. T., *Nonvascular Plants: Form and Function.* Belmont, Calif.: Wadsworth Publishing Company, 1964.

Ebert, J. D., and I. M. Sussex, *Interacting Systems in Development,* 2d ed. New York: Holt, Rinehart and Winston, 1970.

Ray, P. M., *The Living Plant,* 2d ed. New York: Holt, Rinehart and Winston, 1972.

Scagel, R. F., R. J. Bandoni, G. E. Rouse, W. B. Schofield, J. R. Stein, and T. M. C. Taylor, *An Evolutionary Survey of the Plant Kingdom.* Belmont, Calif.: Wadsworth Publishing Company, 1965.

The Fungi

In many ways the fungi parallel the algae. So close are certain members of each group in morphological similarity that many botanists feel that at least some of the fungi had algal ancestors or evolved from the same kinds of ancestors as did some of the algae. On the other hand, there is good evidence that not all fungi had the same ancestry and that at least some had origins among the protozoans. As with the algae, it is probably erroneous to refer to the "Fungi" as a group of naturally related organisms. It would be better to regard them as plants or plantlike organisms grouped together only because of certain superficially similar features. Some professional mycologists (botanists who study fungi) are not convinced that fungi are plants at all! They would go so far as to recognize that these organisms lacking chlorophyll belong to a kingdom separate from the plant and animal kingdoms. But whether the plantlike characteristics of some fungi are the result of a relationship of these organisms to the plant kingdom or are a result of convergence of morphological features, these features are so much like those found in certain algae that the fungi will be considered here. The classification of funguslike plants in Chapter 1 is not intended to be authoritative but is presented simply as one of convenience, organized along the same pattern as the classification system for the algae.

TYPES OF FUNGI Fungi are organisms that are plantlike in many respects, but differ from practically all other plants in that they lack chlorophyll. They are nonvascular plants and are often filamentous in form. More complex body types represent an aggregation of these fungal filaments. Walls of the cells or filaments generally have cellulose, chitin, or both. Somewhere in the life cycle of fungi, spores are produced. Fungi include unicellular forms as well as more complex body types. Life cycles of many fungi parallel those of certain algae, although by no means do all fungi have algal life cycles. As in certain algae and in the land plants, furthermore, many fungi have alternating cytological phases; plants with haploid nuclei succeed those with other chromosome states, and those states in turn may succeed the haploid phase. The relative development of each cytological phase differs in the different fungal groups.

Although it is a distinct possibility that at least some fungi are closely related to algae, the subsequent evolution of the two groups has led to widely divergent paths. Algae, in general, are adapted to an aquatic environment and are dependent upon this environment for the source of nutrients, for raw materials for photosynthesis, and for effecting the reproductive process.

Because of their achlorophyllous nature, fungi have become adapted along different lines. They are dependent upon some substrate for nourishment, and even the aquatic forms, while resembling algae to a great degree, have a completely different mode of obtaining necessary food materials.

Certain fungi have the ability to utilize nutrients from organic matter in their environment. These organic materials may be in solution, or the fungus may play an active part in their degradation, thus making available foods that were originally in the insoluble state. Fungi that exist on nonliving substrates are called *saprophytes.* The source of their nourishment may be such diverse things as a rotting log, a slice of bread, wood, or a piece of cellophane.

Parasites, on the other hand, invade tissues of living organisms and are often responsible for diseases in these organisms. Certain parasitic forms are able to survive on nonliving organic material, while others can exist only when living on another organism.

Most fungi are able to reproduce by both sexual and asexual means (a number of fungi in which the sexual process has not been observed are grouped into a category called "Fungi Imperfecti"). The production of airborne spores is an extremely successful specialization of the fungi. The majority of fungi exist in a terrestrial environment, and spores are distributed in the air. Fungus spores are produced in countless quantities, and are exceedingly small and light. As a result, the air is filled with floating fungal spores that may retain their viability for long periods of time and can germinate in a great variety of places.

This group, which depends upon outside sources of food, has had an extremely diverse evolution. The result is a wide range of body types, reproductive mechanisms, and associations. It is safe to say that no other single major group in the plant kingdom has attained such a great degree of diversification. In certain cases the filaments of fungi have invaded roots of higher plants without seriously affecting the host, and some have evolved such an intricate host-parasite relationship that neither the fungus nor the host can survive without it. One example of the latter is the necessity for orchid embryos to have within their cells fungus filaments that play a part in digesting food materials in the soil during the germination of orchid seeds.

Fungi may also grow in close association with algae to produce *lichens.* The association is probably a parasitic one on the part of the fungus, but not of sufficient severity to harm the algal hosts. Lichens may assume the form of the fungus involved (although there are often new configurations in addition) and reproduce in the same manner as the fungus (Fig. 3-1).

As in the discussion of the algae, not all of the fungal groups will be surveyed here. Certain of the interesting and evolutionarily significant forms will be presented as representative fungal types.

SOME EVOLUTIONARY TRENDS OF ALGALIKE FUNGI

Among the oomycetous fungi are forms that closely parallel the filamentous tendency among the algae. Some of these fungi actually live in the water and reproduce asexually by means of flagellated spores. One order, the water molds, though common in fresh water, will also tolerate water with a small degree of salinity. One family within this order has forms that are branched filaments lacking cross walls except in areas immediately subtending reproductive cells. Numerous haploid nuclei are contained in these nonseptate filaments. Except for the lack of chloroplasts, these fungi have close parallels among the tubular or nonseptate algae.

Saprolegnia represents a fairly primitive type in the family with regard to the asexually reproducing stage. At the terminus of a filament, a cross wall cuts off an elongated *sporangium* (Fig. 3-2A). The contents of the sporangium divide into numerous individual protoplasts, spores, each with a single nucleus. These subdivisions of the sporangium become biflagellated and, when released, swim away as zoospores. Soon flagella are lost, and the spore encysts. Upon germination of the cyst a new flagellated zoospore is released and produces a new filamentous individual.

Sex organs are produced on some of the plants. These are of two types: enlarged cells with a number of spherical, multinucleate female gametes (eggs); and tips of other filaments that develop into male sex organs (Fig. 3-2B). These male sex

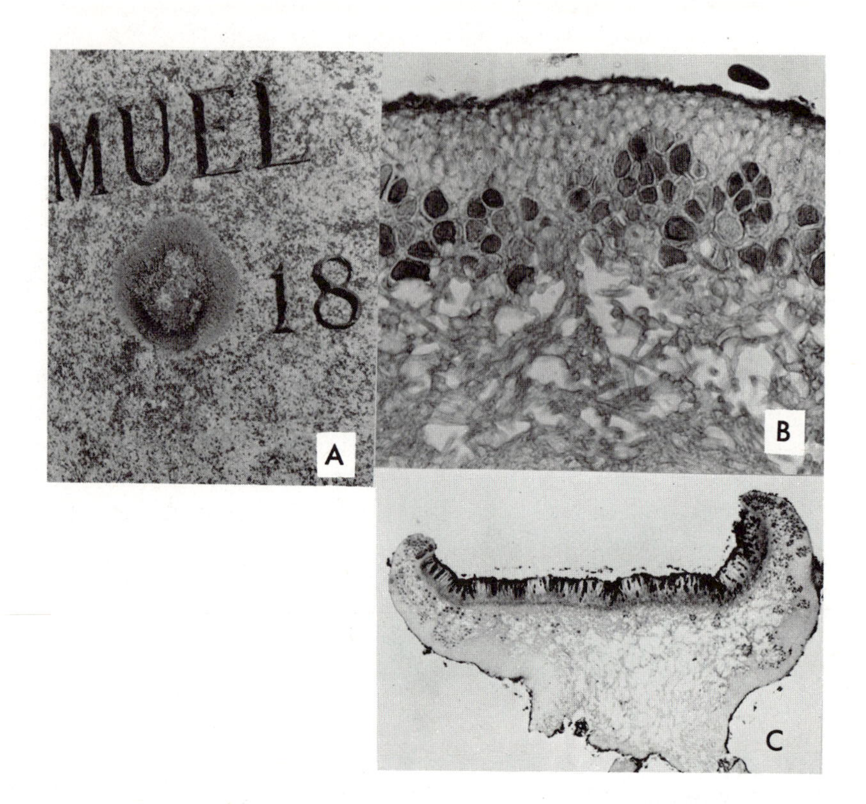

Fig. 3-1 (*A*) Caloplaca, *a crustose lichen about 3 inches in diameter, on a nineteenth-century granite tombstone.* (*Photograph courtesy Conan J. Taylor, O.F.M.*) (*B*) *Section of part of the lichen* Physcia, *showing algal cells* (*darker bodies*) *growing among the fungal filaments.* $\times$*333.* (*C*) *Longitudinal section of the lichen* Physcia, *showing a cup-shaped fruiting body similar to those found among the sac fungi* (*compare with Fig. 3-6*). $\times$*53.*

organs become appressed to the female sex organs and actually penetrate them with short, tubelike processes. Male nuclei eventually fuse with those of the egg. A tube develops from the fertilized egg and forms a sporangium at the tip, releasing motile zoospores. It is suggested that meiosis may occur within the gamete-producing structures; if this is true, then most of the life cycle is in the diploid state. This life cycle is like that of certain algae that are diploid throughout except for the short stage after production of haploid gametes.

Though somewhat simplified, the preceding discussion of the life cycle of *Saprolegnia* serves to give, in an idealized fashion, some indication of the reproductive process of one of the water molds. There may be a number of variations and additional complexities, but the main scheme is as presented here.

In the same family there is an interesting evolutionary sequence in sporangial modification. *Saprolegnia* produces in the sporangium sporelike bodies that have two flagella. The genus *Achlya*, with a life cycle essentially like that of *Saprolegnia*,

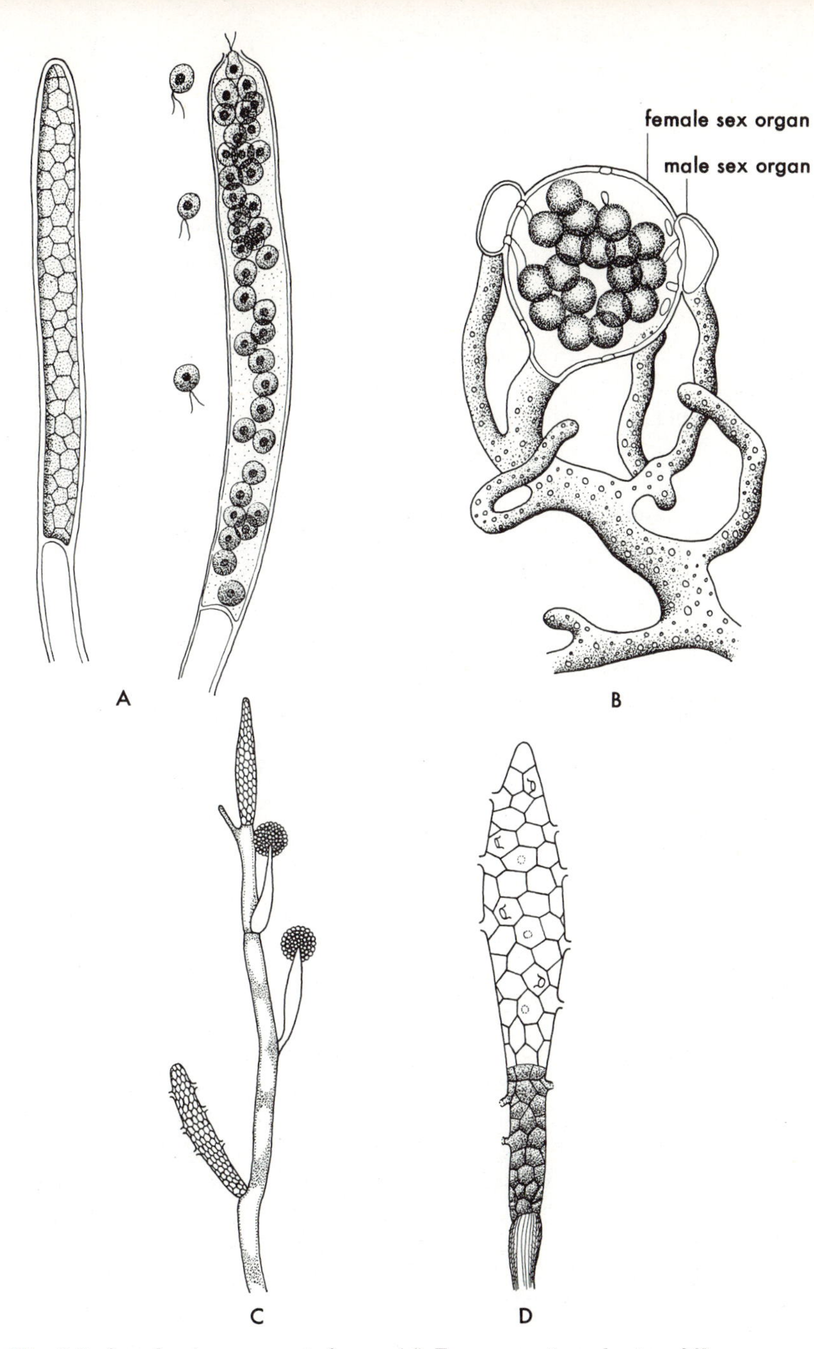

Fig. 3-2 Saprolegnia, *an aquatic fungus. (A) Two sporangia at the tips of filaments, one with divided cellular contents and the other having released motile zoospores. (B) Sexual reproduction, with the male sex organs appressed against the female sex organ containing eggs. [(A) redrawn from William H. Brown,* A Textbook of General Botany, *1925, published by Ginn and Company. (B) Redrawn from* The Plant Kingdom, *by William H. Brown, © Copyright, 1935, by William H. Brown. Used by permission of the publisher, Ginn and Company (Xerox Corporation.)] (C)* Achlya, *with the tip of the filament showing release of contents of the sporangia in a nonmotile form prior to the release of motile spores. (D) Sporangium of* Dictyuchus, *the contents of which have released zoospores through pores. [(C) and (D) redrawn from E. A. Gaumann and C. W. Dodge,* Comparative Morphology of Fungi, *McGraw-Hill Book Company, 1928. By permission of the publishers.]*

differs in that cells that will eventually produce the spores flow out of the terminus of the sporangium and form a tight cluster (Fig. 3-2C). The walls of these spore initials then rupture, each releasing a single biflagellate zoospore that swims away to produce a new individual. In other words, the first motile stage of zoospores is eliminated.

A still further modification is found in *Dictyuchus* of the same family. Here the cells that produce zoospores do not leave the sporangium at all. There are membranes delimiting these spore initials, and when the zoospores are formed within these cells, they are released directly through openings in the sporangial walls (Fig. 3-2D).

Among the water molds there are differences in the sexual processes with respect to the kinds of individual plants involved. Certain of these forms are hermaphroditic; that is, both male and female sex organs are borne on the same plant, or even on the same filament, and the male organs are simply branches arising from the base of the female organ. In other forms (even within a genus that has at least some hermaphroditic species), two individuals are necessary to effect fertilization. Two different mating types, when in proximity, stimulate the production of sex organs, with the male organs formed on one plant and the female ones on the adjacent plant. Because the gametes are of two distinct types it is possible to refer to these two mating types as male and female, unlike the situation in certain algae where the gametes are morphologically similar, requiring reference to the two types of plants as + and − strains.

A different type of life cycle occurs in *Allomyces*, an aquatic member of another division, the chytrids. In some members of the genus there is a regular alternation of phases—one involving the haploid, gamete-producing phase, and the other the diploid, spore-producing phase (Fig. 3-3). Again, the filamentous habit and life history of certain algae are duplicated here. At the tips of the branches of the haploid plant are borne two types of sex organs, an ultimate male gamete-bearing structure and a penultimate female gamete-bearing structure. Gametes are motile, each with a single flagellum; male gametes are smaller than the female. After fertilization a diploid zygote results, and the spore-producing phase of the life cycle is initiated. A diploid filamentous plant is produced by the germinating zygote. On this diploid plant are sporangia; one produces diploid, uniflagellate zoospores that germinate into new diploid plants, and the other produces sporangia in which reduction division occurs. Zoospores produced in the latter are haploid and also uniflagellate. These germinate into haploid, gamete-bearing plants.

In habit and life history the aquatic fungi discussed thus far have much in common with certain algae. In fact they also occur in the same kind of environment. The many features that these aquatic fungi share with the algae may be interpreted as indicating either a close relationship between the two groups or that their origins may have been from the same kind of unicellular ancestor. The close resemblance, it might be argued, agrees in so many features that we may find it difficult to imagine so many coincidental parallelisms. The alternative is to interpret the similarities as an example of convergence influenced by the environment. The filamentous habit and the reproductive mechanisms, including motile spores, may be of considerable

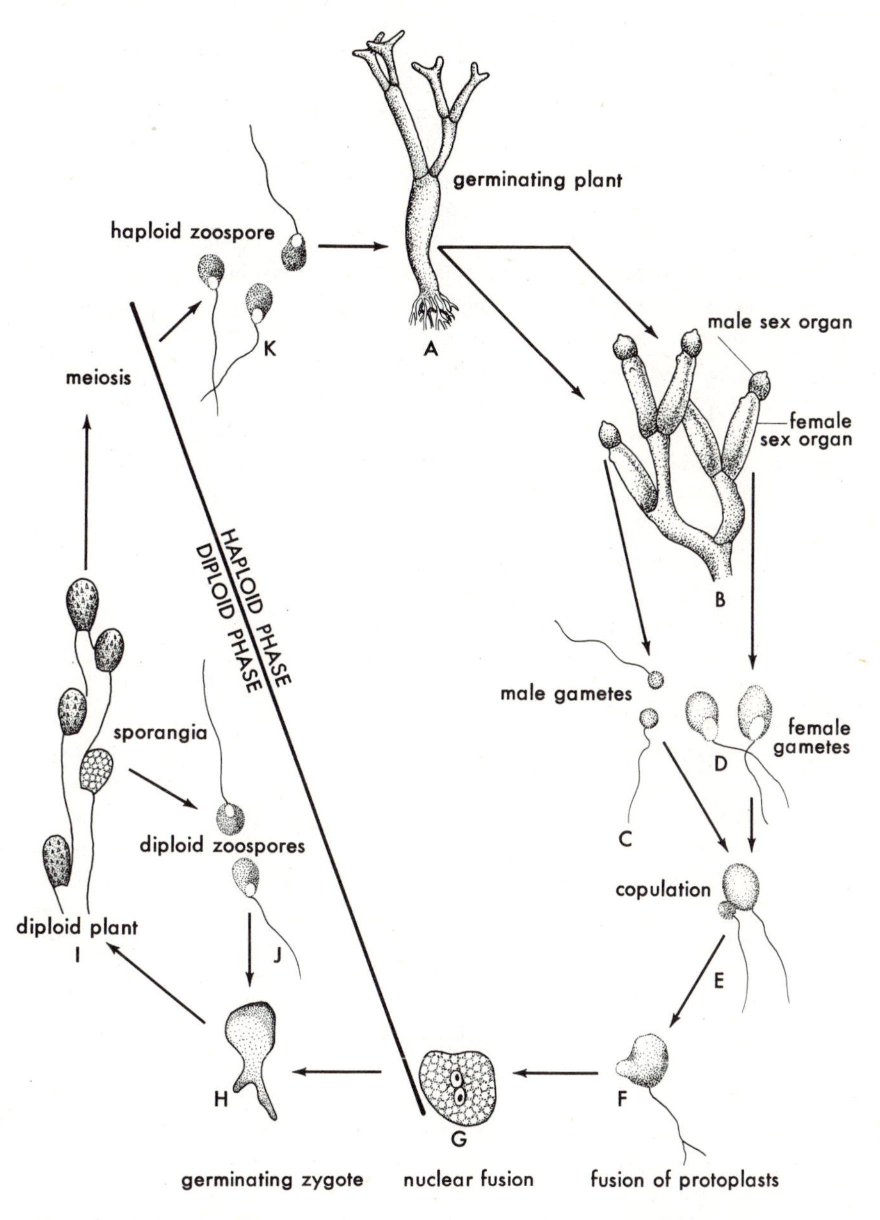

Fig. 3-3 *Life cycle of* Allomyces macrogynus, *showing the alternation of a haploid phase with a diploid phase. (Redrawn from C. J. Alexopoulos,* Introductory Mycology, *2d ed., Wiley, 1962. By permission.)*

survival value in aquatic environments, and these are features that may have been assumed by more than one unrelated group.

Terrestrial Fungi Resemblance of the chytrids and oomycetous fungi to the algae is closest among the aquatic forms. Fungi have become quite successful in terrestrial habitats as well, however, and a number of modifications making these organisms adaptable to the land have resulted in a considerable divergence from the body plan and biology of aquatic forms. In these terrestrial forms, motile spores are absent, and there is great dependence upon small, light, airborne spores. So efficient is this method of distribution that it is difficult to maintain sterile conditions in laboratories, hospitals, and in the home. The air is literally filled with fungal spores that germinate with little difficulty when conditions are suitable. Many of these spores are of algal fungi.

Spores of the common black bread mold *Rhizopus*, a zygomycetous fungus, are especially abundant and easily transported, and for this reason chemicals are often introduced into the dough used for bakery products to inhibit the growth of this and other fungi. *Rhizopus* has a filamentous body plan, consisting of slender filaments, along certain parts of which are borne clusters of shorter, branched filaments that serve to anchor the organism and act as organs of absorption (Fig. 3-4A). Arising from the horizontally creeping threads are erect branches, each terminated by a swollen globose sporangium. Here thousands of black spores are produced and disseminated by air currents; when conditions are appropriate, these spores germinate to produce new organisms. Sexual reproduction occurs when two different strains of the plant come together. Then the ends of two lateral branches of the compatible strains come into contact and a wall cuts off each of the tips of the copulating branches (Fig. 3-4B). The point of contact between the two branches dissolves and in the large cell thus produced the protoplasts of the two branches mix, and the nuclei pair and eventually fuse. The cell in which the two strains of protoplasts mix becomes enlarged and thick-walled, and may undergo a long rest period before germination. The exact time of meiosis is undetermined, but it occurs during the germination process of the zygote, at which time the thick-walled diploid cell ruptures and forms a stalked sporangium similar to those on the vegetative thalli.

An impressive laboratory demonstration that shows the necessity of two strains for sexual reproduction in *Rhizopus* involves culturing two compatible strains on a dish of agar. Where the filaments of the two strains come in contact, copulation occurs, as evidenced by a line of conspicuous black spheres, the thick-walled zygotes, through the middle of the dish along the line of mixing.

Rhizopus has a haploid chromosome condition practically throughout its entire life history. The only diploid phase occurs after fertilization and involves only one cell, the zygote.

Many more members of the chytrids, oomycetous fungi, and zygomycetous fungi are known, but the foregoing examples represent some of the typical, more common forms. Parallel structures with algae are quite evident, and in certain cases

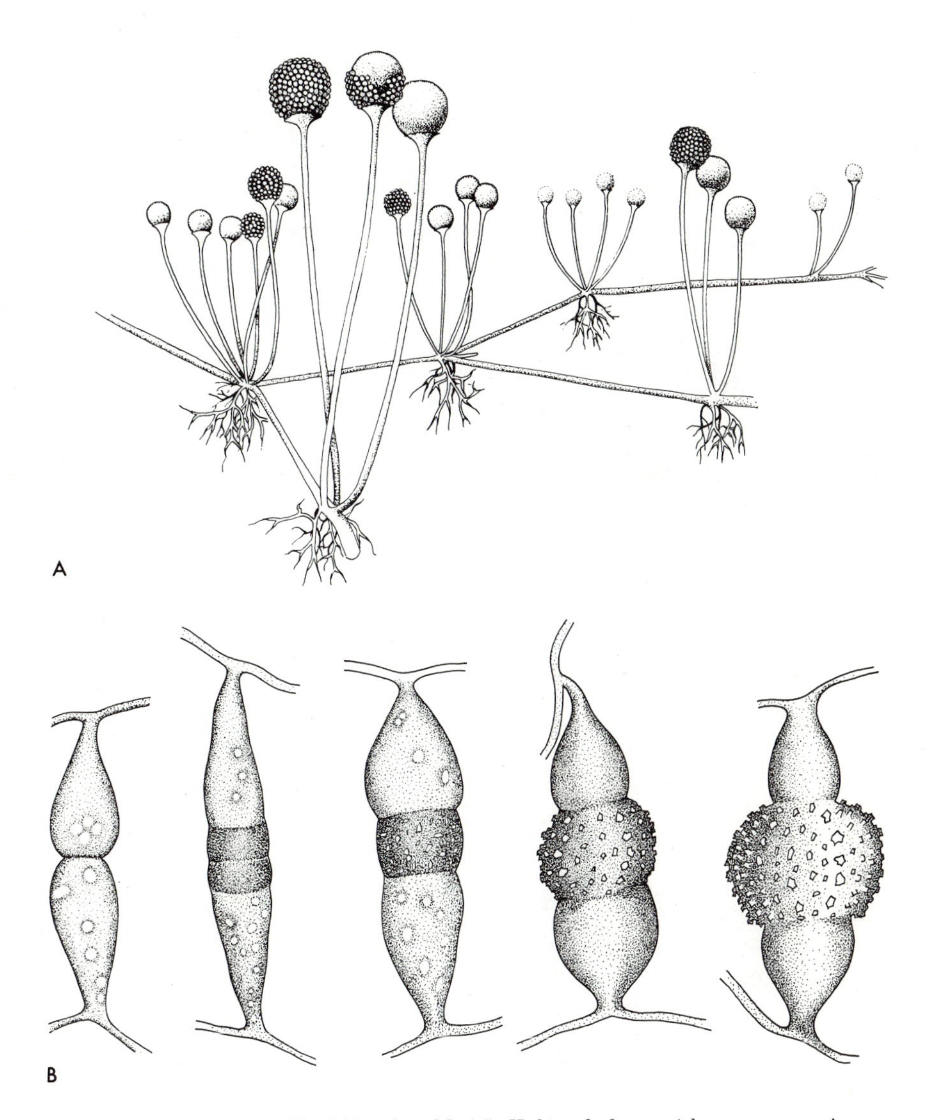

Fig. 3-4 Rhizopus, *the black bread mold. (A) Habit of plant, with erect sporangia. [Redrawn from* The Plant Kingdom, *by William H. Brown,* © Copyright, 1935, *by William H. Brown. Used by permission of the publisher, Ginn and Company (Xerox Corporation.)] (B) Stages in sexual fusion, with branches of two adjacent filaments meeting at their tips to form the large, diploid zygote. (Redrawn from William H. Brown,* A Textbook of General Botany, *1925, published by Ginn and Company.)*

they are practically identical. Modifications in the body habit of these fungi might be considered analogous with similar development in the algal groups, the principal divisions of which are not necessarily closely related. These fungi are of special interest in showing a possible evolution from forms that were essentially aquatic to forms that are extremely specialized and adapted for an existence under terrestrial conditions.

THE SAC FUNGI The sac fungi characteristically have septate filaments as opposed to the generally nonseptate nature of the forms discussed earlier. Furthermore, some time after the sexual process there is formed a characteristic elongated saclike structure, an *ascus* (plural, *asci*), in which reduction division occurs and in which there are typically (but not universally) eight haploid spores, the *ascospores.* There is a considerable range in body form and size among the sac fungi, which vary from minute, unicellular forms to massive, elaborate types, but none of the varieties produces flagellated cells.

The yeasts (Fig. 3-5) include forms that are unicellular and are among the simplest of sac fungi. Mycologists are not agreed about whether the simplicity of the unicellular habit is a primitive feature or whether it is derived by reduction from a multicellular form.

Certain yeasts are simple, ellipsoid cells, each with a single nucleus. Vegetative reproduction is effected by a budding of the original cell, with one of the two nuclei formed after division migrating into the lateral bud. Continued vegetative budding may produce chains of new cells that are held together loosely and that eventually break apart into individual cells. Sexual reproduction may occur when two adjacent yeast cells produce short processes and the contents of each process fuse, producing a diploid zygote nucleus. The cell with the fused contents functions as an ascus, with meiosis producing four haploid cells and a further, mitotic, division occurring to form eight cells, the ascospores. Each spore then produces a new individual. In this simple type of life cycle, the diploid phase is only a small part of the entire sequence of events.

Other yeasts are more complex in their reproductive pattern. Haploid cells may bud vegetatively for some time. Eventually two adjacent cells undergo copulation to produce a diploid cell. This diploid individual may then continue to bud vegetatively, with meiosis occurring later. Four haploid cells (ascospores) are formed after meiosis. Here there is an actual alternation of haploid and diploid phases.

Yeasts are not typical of the usual type of life cycles among the sac fungi. More "normal" are the multicellular forms that produce a rather extensive haploid body. On specialized branches the plant produces small, nonmotile spores that are disseminated by air currents and produce new individuals. Sexual reproduction involves a short lateral multinucleate branch, the male sex organ, and another short multinucleate branch with a swollen base and a slender terminal extension, the female sex organ (Fig. 3-6). The tip of the male organ comes in contact with the tip

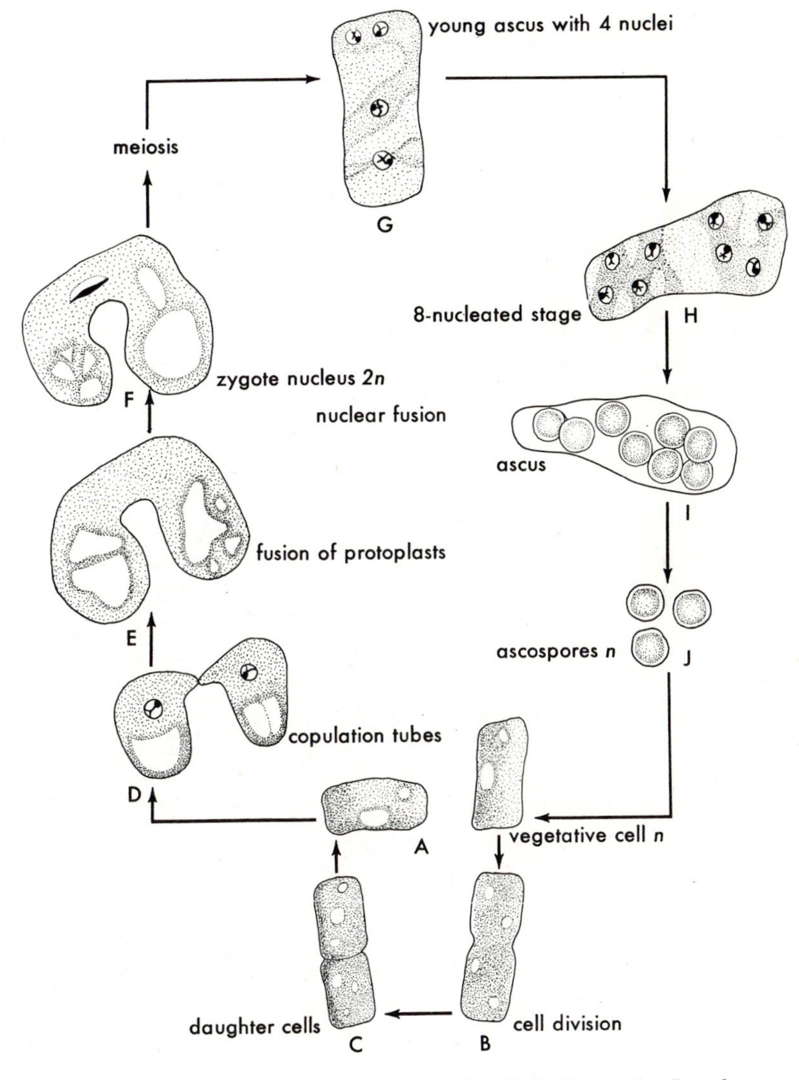

Fig. 3-5 *Life cycle of one of the yeasts. (Redrawn from C. J. Alexopoulos,* Introductory Mycology, *2d ed., Wiley, 1962. By permission.)*

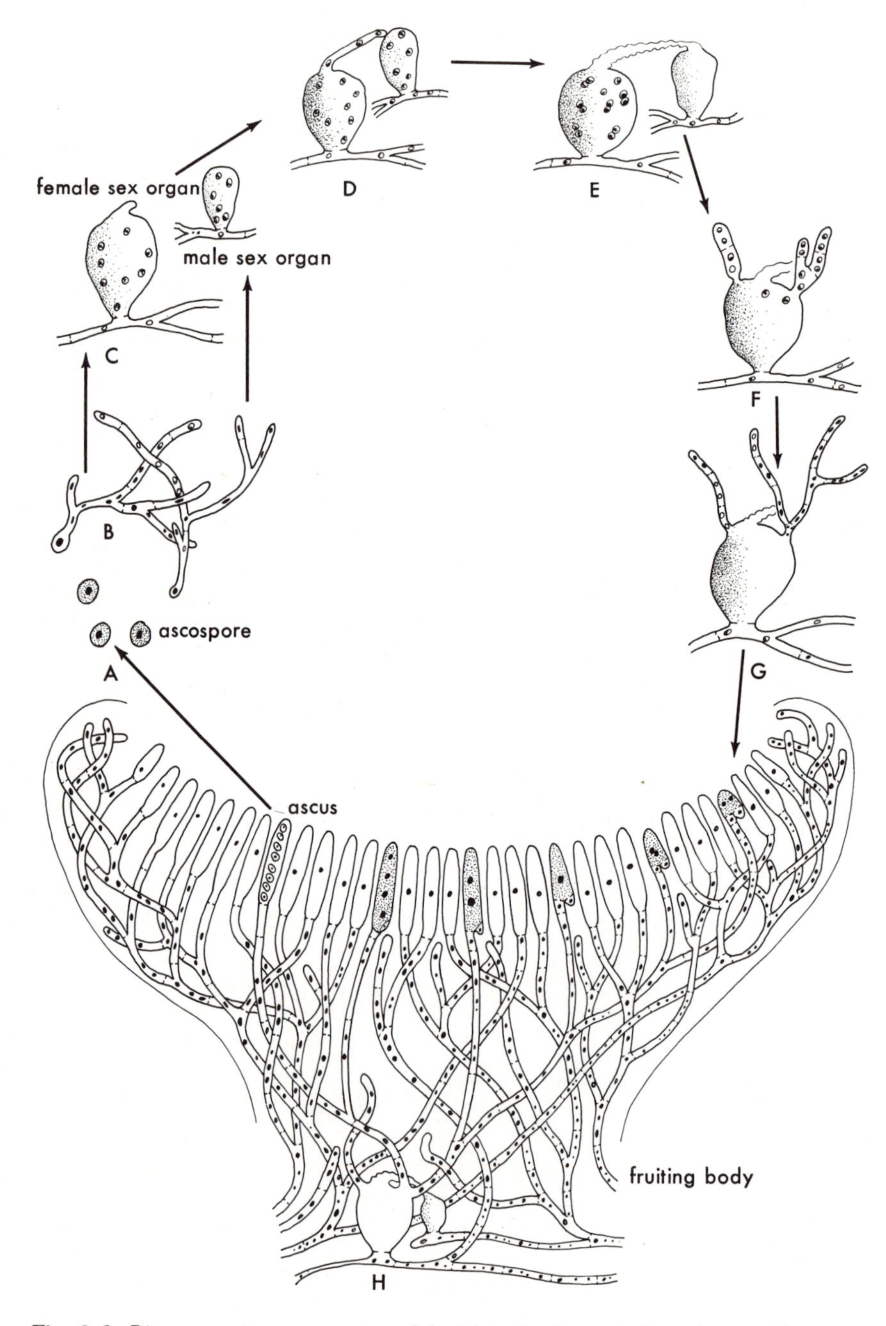

Fig. 3-6 *Diagrammatic representation of the life cycle of a typical sac fungus. (Figure "Life cycle of a typical ascomycete" redrawn from* Introductory Botany *by Arthur Cronquist. Copyright © 1961, 1971 by Arthur Cronquist. Reprinted by permission of Harper & Row, Publishers.)*

of the female structure, with subsequent migration of male nuclei into the female body. The nuclei then pair, each male nucleus with a female nucleus. From the female sex organ a number of lateral filaments develop, into which the paired nuclei move. An ascus is formed at the tips of some of the filaments, and in each of these, two nuclei unite to initiate a very brief diploid phase, terminating the stage in the life cycle in which the nuclei are simply paired. Meiosis occurs within the ascus, generally followed by another division to produce eight haploid cells, the ascospores.

In the sac fungi the asci are often grouped into elaborate and sometimes colorful fruiting structures; some of the fleshy woodland fungi are sac fungi with fruiting bodies that may be in the form of cups, spheres, urns, flask-shaped structures, or a variety of other shapes. The tasty morsel is a sac fungus with a large, fleshy fruiting body.

Cytologically, the principal part of the life cycle of a sac fungus is in the haploid chromosome condition. The diploid phase involves only a single cell. A new phase, however, that in which male and female nuclei remain paired but unfused, has been introduced. Although this phase is longer than the diploid stage, it is still a relatively small part of the entire life cycle.

THE CYTOLOGICAL STORY IN THE CLUB FUNGI

The club fungi are an extremely successful group of heterogeneous fungi. These forms have septate filaments (as in the sac fungi), and the special spore-producing structure is often a club-shaped cell that typically produces four haploid spores on the surface. As in the sac fungi, no flagellated cells are formed.

Among the club fungi, germination of a haploid spore typically forms a filament with uninucleate cells. Subsequently there is the beginning of a sexual process that does not involve conspicuous sexual structures. The tips of two filaments from different individuals may simply come together and their terminal cells unite. Two associated nuclei (one from each cell) thus come in contact, but they do not unite. After many cell and nuclear divisions an extensive mass of filaments is produced in which each of the individual cells has paired nuclei. A very significant proportion of the life cycle of a club fungus involves this phase with paired nuclei.

There are other modifications of methods by which two mating types come together. Some forms produce small haploid cells (called *spermatia*) that come in contact with a haploid filament of a compatible mating type; this initiates the phase with paired nuclei.

Development of the plant in this cytological state may result in the production of the part of the life cycle that is the most conspicuous. Fruiting bodies of mushrooms and puffballs, for example, have filaments with paired nuclei. However, the haploid phase is generally quite insignificant and is never extensively developed.

Eventually, club-shaped terminal cells of the filaments with paired nuclei are cut off. In each terminal club-shaped cell a pair of nuclei fuse, producing the short diploid phase. Subsequently, meiosis occurs in each terminal cell, and the haploid nuclei migrate into four spores that project from the surface.

Although no relationship is implied, there seems to be among the club fungi an elaboration of a cytological "theme" evident in the sac fungi. Practically all of the life cycle of a sac fungus is in the haploid state; both the filaments and asexual spores have haploid nuclei. After copulation, however, nuclei from two mating types pair, but do not fuse. Subsequent growth of the female sex organ produces a small portion of the life cycle that is multicellular, with each cell containing a pair of nuclei. This phase with paired nuclei in lacking among the chytrids, oomycetous fungi and zygomycetous fungi, but is not as well developed in the sac fungi as it is among the club fungi.

In the club fungi, the haploid phase is a very small part of the life cycle. After the pairing of two mating types, an extensive filamentous body is produced in which each cell has a pair of nuclei. Only in the terminal club-shaped cell formed at time of fruiting is the diploid stage realized, before meiosis (just prior to the production of spores) reintroduces the haploid cytological condition.

PARASITISM IN THE CLUB FUNGI

Many of the club fungi (as well as members of the other fungal divisions) are extremely specialized and well-adapted parasites. The host-parasite relationship is often complex, and there are many instances in which two host plants are necessary for the survival of the parasite. The common wheat rust fungus, for example, infects leaves and stems of wheat plants during much of its life history. The haploid spores produced after the sexual phase, however, infect the barberry plant. The fungus grows on this host for a time, after which other kinds of spores are produced. These fungus spores on the barberry reinfect the wheat. Although both hosts, wheat and barberry, are generally essential for the completion of the entire life history of the wheat rust fungus, the barberry may be bypassed if some of the asexually produced spores from the fungus on the wheat overwinter and reinfect the wheat plants the following spring.

The rusts and smuts, common club fungus diseases, demonstrate an exceedingly intricate host-parasite relationship. So well established is this relationship that, so far as we know, the fungus involved in each instance is able to survive only on the host to which it has become adapted. It is extremely difficult to raise some of these parasitic fungi in laboratory culture. When techniques of culturing such forms are perfected, factors involved in their growth will be better understood, and more effective control measures may be developed.

FUNGI AS EXPERIMENTAL ORGANISMS

Some of the classical work on biochemical aspects of genetics has been done utilizing certain fungi. One aspect of many of the fungi used that makes them quite suitable for genetic investigations is the haploid nature of the plant for a considerable part of the life cycle. It is possible, therefore, to determine the effects of a single gene, which conceivably could be masked by its allele in a diploid organism. The easy culture of fungi in the laboratory and their rapid life cycles also contribute to their desirability as experimental organisms. Bacteria, too, serve as organisms suitable for experimental studies, again primarily in genetic investigations.

BIOLOGY OF ASSOCIATIONS BETWEEN FUNGI AND OTHER ORGANISMS

A relationship between a fungus and another kind of organism that eventually results in a parasitic association is affected by the evolution of both host and parasite. A mutation in a host may make the parasite nonadapted for the change, and unless some corresponding selective change occurs in the parasite, it is no longer adapted for survival on the host. This principle is of significance in attempts to combat plant diseases. In many instances it is virtually impossible to control the fungus (or virus, in certain plant diseases) by chemical means, and the surest way to eradicate the parasite is to take advantage of occasional mutations in the host that produce in it a certain degree of resistance to the parasite. This, of course, cannot be a terminal process because at some later stage the parasite, after undergoing some selectively advantageous mutations, may again be fit to attack the host.

The origins and subsequent evolution of associations between fungi and organisms that actually derive benefit from the presence of the invader are even more intricately interconnected. The establishment of so precise a relationship between host and fungus could not have occurred as a result of only a single mutational change, but more likely occurred after a series of changes. Furthermore we must presume that there were almost simultaneous changes in both host and fungus, the combination of which must have been suitably adapted.

The lichen association is a complicated example of this mutual adaptation. If the isolated fungus and alga involved in a particular kind of lichen are grown together, the fungus produces a substance that kills the algal cells. In some instances the fungus itself may invade the algal cells, causing their death. There has been some success in synthesizing lichens in culture, but only when both components are grown on a substrate that normally does not support growth of either individual. In such a case it seems probable that each component provides some factor that is not available to the other but is necessary for the other's survival. There is still a great deal to be learned about the biology of lichens, and much of this information could probably contribute to our understanding of evolutionary principles in general.

FURTHER READING

Ahmadjian, V., "The Fungi of Lichens," *Scientific American*, vol. 208 (February 1963), pp. 122–132.

Alexopoulos, C. J., *Introductory Mycology*, 2d ed. New York: Wiley, 1962.

Bold, H. C., *Morphology of Plants.* 3d ed. New York: Harper & Row, 1973.

Christensen, C. M., *The Molds and Man.* Minneapolis: University of Minnesota Press, 1961.

Levine, R. P., *Genetics,* 2d ed. New York: Holt, Rinehart and Winston, 1968.

Scagel, R. F., R. J. Bandoni, G. E. Rouse, W. B. Schofield, J. R. Stein, and T. M. C. Taylor, *An Evolutionary Survey of the Plant Kingdom.* Belmont, Calif.: Wadsworth Publishing Company, 1965.

Webster, J., *Introduction to Fungi.* London: Cambridge University Press, 1970.

chapter **4**

Origin and Evolution of Land Vascular Plants

One of the most significant steps in the evolution of the plant kingdom was the migration of green plants from an aquatic environment to a terrestrial one. Whether this migration occurred only once or more than once is not known, but it is generally agreed by plant morphologists, though there is still no way to prove this assumption, that the ancestral aquatic plants that led to the land forms were members of the green algae.

THE GREEN ALGAE AS POSSIBLE ANCESTORS

If we inspect the characteristics of land vascular plants, there seems to be no other extant algal group that may be considered a likely candidate for the ancestor. Land plants, for example, are chlorophyllous (they contain chlorophylls *a* and *b*), and the combination and proportion of green and nongreen pigments (carotene and xanthophyll) in the plastids approximate those of the green algae. Furthermore, the usual photosynthate is a simple starch in both the green algae and the green land plants. On the basis of pigments and metabolites, therefore, the green algae have the most to favor them.

But in spite of the close correspondence of chemical factors, are the structural features of both groups close enough? Land plants almost universally have a bulky, three-dimensional body plan; most green algae do not. There are some forms (for example, *Fritschiella*, Fig. 2-5E), among the green algae that tend to build up a three-dimensional, fleshy plant body. Green land plants have an oogamous type of sexual reproduction; that is, a nonmotile egg is fertilized by a small sperm cell that must somehow be transferred to the egg. Since sexual reproduction in the green algae, as we have seen, ranges from isogamy to specialized oogamy, there is no problem with that feature. Finally, land plants have an alternation of cytological phases; that is, a diploid, spore-producing phase is always followed by a haploid, gamete-producing phase. Again, although among the green algae there are many kinds of life cycles, many forms have just such an alternation of haploid and diploid phases.

Although certain of the other algal groups would seem to qualify as land plant ancestors on the basis of many features, only among the green algae is found the proper combination of prerequisites for a land plant ancestor. The fossil record, unfortunately, is of little help at the present, because even if a possible ancestral algal type were found, no means are now available to assist in determining to what group of algae it belonged. Many times, much of the biochemical nature of a fossil organism is preserved, but not enough to allow us to determine whether an ancient alga was green or not.

Paleobotanists, however, are offering positive contributions toward a solution of the problem. Even though it is not possible to recognize an ancestral algal form, the form of certain early and primitive fossil vascular plants has much in common with certain algal body plans. Furthermore, the fossil record offers evidence as to when land plants first appeared on the earth's surface. By the Devonian Period a wide variety of land plants existed, among them relatives of the club mosses, horsetails, ferns, and other primitive vascular plants. Before the Devonian Period, however, there are only meager remains of vascular plants, suggesting that there may have been a sudden flourishing of land plants toward the end of the Silurian Period, a time about 450 million years ago.

PROBLEMS FACED BY EARLIEST LAND PLANTS

The magnitude of the evolutionary steps from the sea to the land can be realized only when one considers the kinds of modifications necessary if a plant that normally lived in an aqueous environment is to survive on land. The most formidable problems are those related to the availability of water. An aquatic plant such as a green alga has a virtually limitless supply of water if it is living submerged in water. Water and dissolved minerals enter the plant from practically every part of the plant body; they simply diffuse through the outer membranes of the cells exposed to the water. A plant on land, on the other hand, has no such ready supply of water and minerals. The only reliable source is from the soil in which the plant is growing, but since the soil has little or no water close to the surface, the source must be some distance below. Thus it is obvious that to be successful on land a plant must have some way of perpetually obtaining water and dissolved minerals from the soil and conducting them throughout the plant.

Once water has been obtained, it must be conserved. Since algae have no means of conservation, placing an alga on the land would result in its immediate desiccation. The dry air would cause an evaporation of the water from the cell surfaces, and within a matter of minutes most green algae would be dried and dead. To protect a land plant from the harsh drying land environment, especially during times when little or no water is available, it became necessary to develop some means for preventing loss of water from aerial parts of the plant in excess of that necessary for the carrying on of certain functions in the plant.

Another problem involves support. Submerged algae are buoyed up by the surrounding aqueous medium. A similar type of plant placed on land would simply be flattened out against the ground. Land plants, therefore, needed to evolve a more rigid type of body.

These major problems must be taken into account in attempts to demonstrate how the probable green algal ancestors, over a long period of time, became modified into land-dwelling vascular plants.

EVOLUTION OF ROOTS A root system, which is a feature of land plants that is not found among any of the algae, is a specialization that is obviously the result of an adaptation to the kind of environment in which the plant found itself. Many algae have "holdfasts," that is, means of attaching the plant body to some rigid structure at the bottom. These, however, are not absorbing organs as are the roots of higher plants, nor do they serve the same kind of anchoring function that a stout root system provides for the sometimes massive, "top-heavy" aerial portions of vascular plants. Furthermore, with the development of roots, land plants had a ready means of extracting water and minerals from the soil and drawing them into the plant itself.

Roots in the plant kingdom obviously originated later than the ancestral algae, but they must have been derived from some kind of ancestral condition that was present among the algae. In the early part of this century the French botanist Lignier published his views concerning the origin of roots in vascular plants. He proposed that the green alga type of plant body that preceded land plants was a three-dimensional, dichotomously branched system (Fig. 4-1A). In the evolution of roots, one of

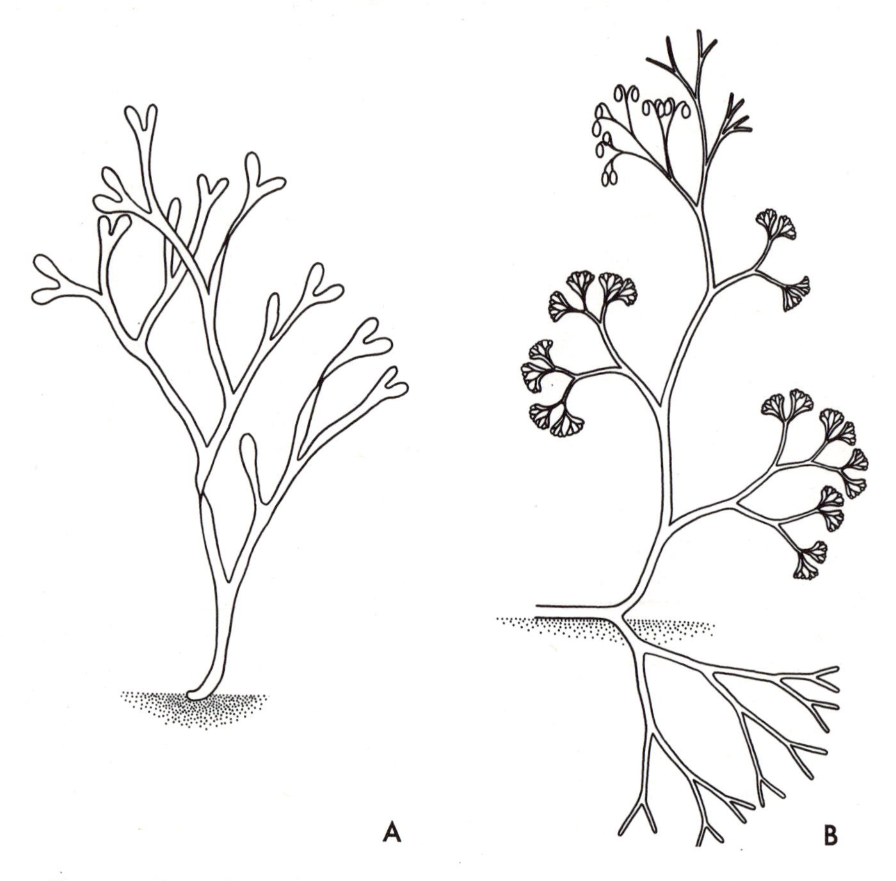

Fig. 4-1 *Origin of a land vascular plant body according to the hypothesis of Lignier. (A) Dichotomizing green algal ancestor. (B) Simple vascular plant.*

the lateral branch systems bent over, eventually penetrating the soil, and branched underground (Fig. 4-1B). In other words, the root system would be considered homologous with part of the aerial branching system. If such a hypothesis were true, one would expect to find in the fossil record some primitive land plants that did not have true roots, but instead had an anchoring and absorbing system that was only a slightly modified branch system. In fact, dozens of such kinds of primitive land plants are known from the Devonian Period (for example, Fig. 4-2A). In these forms, the plant consists of a horizontal stem system, some of the branches of which appear to have penetrated the ground, others of which ascended and became photosynthetic portions of the plant. In later geological periods the root, in an environment quite different from that of the rest of the plant, became considerably more specialized and progressively more distinct structurally. Hence in most vascular plants living at the present time, roots have a number of characteristics that differ markedly from those of stems.

Thus roots were a development of land vascular plants that served the dual functions of anchoring the plant to the ground and absorbing from the soil the water and dissolved minerals that no longer enveloped the entire plant body.

THE CUTICLE Even though absorption of water from the soil makes it available to the aerial parts of the plant, the relatively dry atmosphere on the land would bring about a rapid evaporation of most of

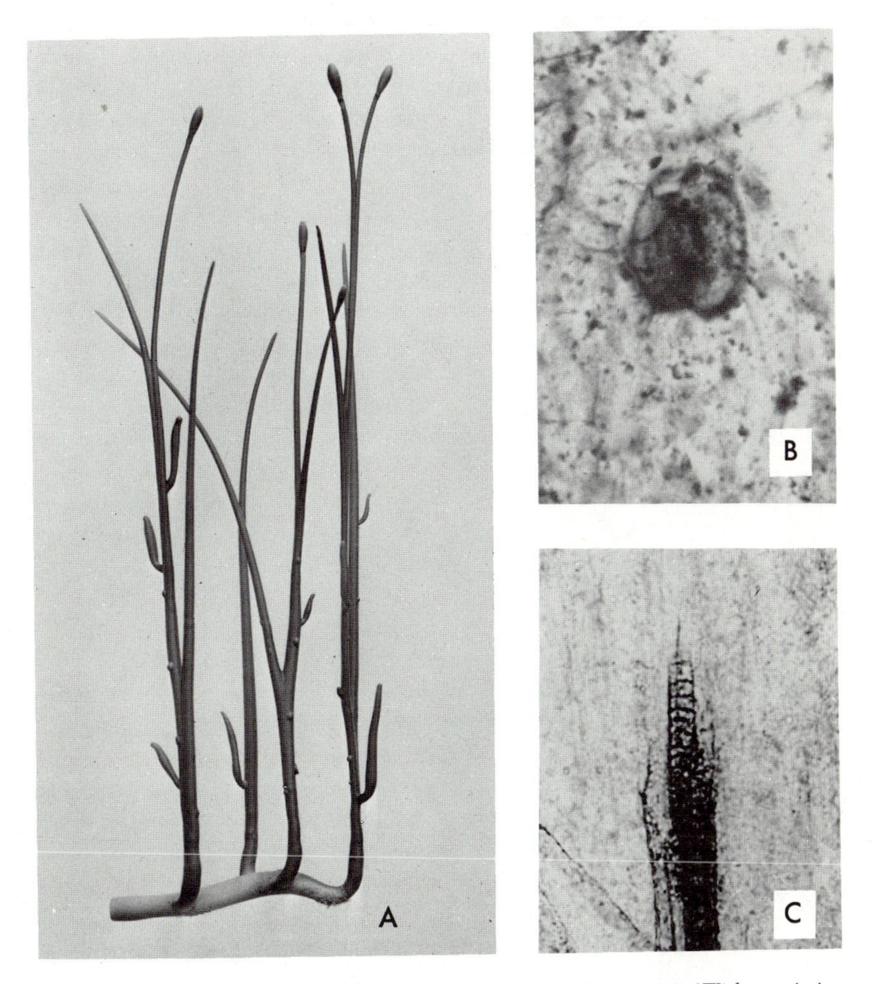

Fig. 4-2 Rhynia gwynne-vaughani. (*A*) *Reconstruction of plant.* $\times$*0.8.* (*With permission of the Chicago Museum of Natural History.*) (*B*) *Longitudinal section of the epidermis, showing a stoma and its accompanying guard cells.* $\times$*240.* (*C*) *Longitudinal section of the stele, with annular tracheids.* $\times$*250.*

the water in the aerial parts if there were no means of retaining it within the cells during periods of water stress. This problem is solved by a feature typical of all aerial portions of vascular plants on land, a thin layer of waxy material, the *cuticle,* which forms a waterproofing layer at the surfaces. It is not a layer of cells, but simply a thin film spread over the outer epidermal structures of plants. This cuticle is extremely effective in helping to retain water within the organs that are exposed to the air.

A considerable amount of water loss occurs from leaves of vascular plants through pores, called *stomata* (singular, *stoma*), which are open for the exchange of gases in and out of leaves. This evaporation of water from leaves directly influences uptake of water from the soil by roots; when water absorption lags behind water loss, the stomata close, thus preventing additional evaporation. At such times the cuticle seals off the rest of the leaf surface, and water balance within the plant is restored.

Cuticles are extremely resistant to decay. Many times when all structural features of fossil plant parts are almost obliterated, the cuticle is intact and can be studied to give information about the epidermis of various plant organs. Spores and pollen grains too have an external substance similar to a cuticle that allows these small bodies to be preserved in sediments and studied by paleobotanists in the special branch of their subject called *palynology.*

Plants with cuticles are extremely old geologically, and it is possible to make certain suggestions about the environments in which these plants existed. Although it cannot be stated with certainty that because a plant had a cuticle it must have been a land plant, it is quite safe to suggest that at some time in the life cycle of that plant the parts with cuticles must have been exposed to the air. It would seem most likely that the ancestors of the land plants must at some time in evolutionary history have occupied a position along the tidal zones where they were alternately dried and moistened again.

ORIGIN OF CONDUCTING SYSTEMS

We now know that a newly evolved land plant had a root system that anchored the plant to the ground and served as an absorbing organ, and a waterproofing cuticle that covered the aerial parts to conserve water. If the aerial parts of the plant were some distance from the ground, however, some means of transporting the water and dissolved salts to them had to be developed. A specialized conducting system became the means of transport of soluble substances and water from the root system to all other parts of the plant. It was the evolution of such a system that enabled land plants to attain such large sizes. Furthermore, this conducting system was the means of transporting foods manufactured in the aerial green parts of the plant to all other, nonphotosynthetic regions.

It is true that certain land plants (for example, liverworts and mosses) have no conducting systems such as those found among higher land plants. In practically all cases, however, these plants are rather small, and materials do not have great distances to travel. Furthermore these plants very frequently grow in rather moist habitats, where much water can be absorbed through the surfaces of the aerial parts of the plant. Although some of these bryophytic plants grow in somewhat drier regions, they generally do not grow actively during prolonged dry spells. Instead they take on a shriveled and dried appearance until they are moistened again; then they begin active growth, taking on a lush green and healthy appearance. In other bryophytes there are some indications of a conducting system in the main axis of the aerial part of the plant. This conducting system is not exactly homologous with that in the vascular plants because the conducting cells are quite different in structure. These specialized conducting cells in bryophytes have the same function, however.

In the vascular plants the first conducting system was extremely simple, consisting in most cases of a very slender solid strand of elongated cells in the center of the stem axis. Such simple conducting strands appear in the earliest fossil vascular plants from Silurian and Devonian rocks. The most primitive water- and mineral-conducting cells differ little from the ordinary thin-walled cells in the rest of the plant. More complicated and elaborate conducting systems evolved during later geologic ages. Particulars concerning the evolution of conducting systems appear on later pages.

At present the conducting systems in most vascular plants are quite complex, and in trees the conducting system makes up practically the entire bulk of the roots, stems, and branches. With more elaborate conducting systems, plants gained the ability to grow higher and have distal parts considerably removed from the source of water. Concurrently a more rigid supporting system developed.

CHANGES THAT ALLOW LAND PLANT SURVIVAL

Aerial parts of land plants may extend for some distance from the ground. A vascular system moves water and dissolved materials throughout the plant, a root system anchors the plant, and a cuticle prevents excessive water loss from the surface exposed to the air. Because photosynthesis occurs in cells within the leaf (most of the epidermal cells lack plastids), some provision had to be made for the easy passage of gases through the plant surface to allow entrance of air with its carbon dioxide. Land plants developed specialized stomata on leaves and stems of aerial systems. Carbon dioxide enters the leaf through these stomata and comes into proximity with the photosynthesizing cells.

Finally, even though stem and leaf surfaces are covered by a cuticle, it is a familiar fact that many vascular plants, especially woody ones, increase in diameter as more conducting tissues are produced internally; this increase in bulk of conducting tissues naturally results in the building up of internal pressures that eventually burst the outer tissues and cuticle with them. It is at this stage that cork cells are produced at the outer part of stems. This corky layer becomes the outer limiting layer. Cells of cork at maturity are nonliving and have thick, waterproof walls. The waterproof nature of cork cells is easily noted when one considers that ordinary bottle corks are made from the corky layer of a certain species of oak tree.

It should now be obvious that a number of structural modifications were involved in the evolution of a plant type such as an aquatic green alga to one such as a woody plant. These changes, which must have occurred throughout a long period of time, allow survival of land plants in an environment that is harsher and less constant than the environment in which most algae are situated. The fossil record has provided evidence for certain of these modifications, and botanists are able to verify their evolutionary hypotheses with definite and indisputable evidence.

ALTERNATION OF CYTOLOGICAL PHASES IN LAND VASCULAR PLANTS

It was mentioned earlier that land plants typically have an alternation of phases in their life cycles. There is a rhythmic change from a diploid phase that produces only spores to a haploid phase that is the sexually reproducing part of the plant life cycle. In some land plants (for example, ferns) these two phases are quite distinct and are represented by two independent green plants. In other plants (for example, flowering plants) only one phase, the diploid phase, is conspicuous; the other is extremely reduced and not readily visible. On the other hand, in the Bryophyta the haploid, sexual phase is the dominant one, with the diploid, spore-producing phase reduced and actually dependent upon the haploid phase. This alternation of phases, represented both by chromosomal change and functional differences, is often called *alternation of generations.* Because both phases are actually parts of one and the same life cycle, perhaps it is more appropriate to refer to this alternation as one of phases or stages rather than of generations.

Our brief survey of the algae showed that among them are some forms that have a haploid phase for practically the entire life cycle. Only when two gametes fuse is there a diploid phase, and this consists of only one cell, the zygote, which divides meiotically upon germinating to restore the haploid phase. Other algae have both haploid and diploid phases, often equally developed. In these forms a haploid plant produces only gametes that fuse to initiate the diploid phase. At germination the zygote produces a multicellular diploid body by a series of mitotic divisions that produce spores at maturity. These spores are the results of a meiotic process and are haploid; upon germinating they produce a new haploid sexual phase again.

Land plant alternation of phases is very much like the latter kind of algal life cycle. In no known land plants, however, are the two phases identical in structure, as they are among some of the algae that exhibit this same kind of alternation. The origin of the differences in the two stages of the land plants has been the source of considerable speculation among many botanists. Some regard this difference as the result of dissimilar environmental conditions for each phase, both of which were identical originally, resulting in different structural modifications. Others suggest that the type of algal ancestor that gave rise to the land plants was one with only a haploid phase (except for the diploid zygote) and that the diploid part of the life cycle was the result of a delay of the time of meiosis until after a number of mitotic divisions, starting with zygote germination, had occurred. This delay in meiosis would be responsible for the production of a multicellular diploid phase before the event of spore production, at which time meiosis of the spore-producing cells occurred. The theory suggesting an original alternation of two, probably identical, phases in the ancestral alga is called the "homologous" theory of alternation. The theory proposing a later origin of the diploid phase is called the "antithetic" or "interpolation" theory (in other words, a new phase, the diploid one, is "interpolated" into the haploid life cycle).

EVOLUTION OF THE SPOROPHYTE

As mentioned in the preceding section, in practically all vascular plants the diploid, sporophyte stage is the more conspicuous phase of the life cycle, and the haploid, gametophyte stage is much less prominent. In fact the part of the vascular plant life cycle that we see when we look about us is the sporophyte. An elm tree, a maize plant, a spruce tree, a clubmoss, or a fern, for example, are evident to us from the sporophyte stage. All of these plants, as the layman knows them, are the spore-producing parts of the life cycles. Because the sporophyte is so familiar to everyone and because it is the conspicuous part of the life cycle, considerable attention will be devoted to its evolution; discussion of the evolution of the gametophyte stage will be postponed to the section on reproduction in land plants.

Simplest Vascular Plants

If given the task of trying to imagine the simplest kind of vascular plant, one that most resembled a possible green algal precursor, we might conjure up the image of a slender green branching axis, probably without appendages or a root system. It was just such a simple kind of plant that Lignier envisioned as representing the most primitive vascular plant. If this idea is correct, it should be possible to prove it from the fossil record. In the rocks there are, indeed, just such simple land plants known from the late Silurian and Devonian Periods. One of these, the well-known genus *Rhynia,* described from the Devonian of Aberdeenshire, Scotland, is nothing but a naked branching axis (Fig. 4-2A). It is leafless and rootless, and the branching is dichotomous. Part of the stem system is horizontal, and tufts of slender hairlike extensions (*not* roots) served to absorb water and minerals from the soil. The aerial branches extend upward, and some of the stems are terminated by small, swollen spore sacs or sporangia.

The absence of leaves would suggest that the entire aerial system was photosynthetic. There is no way of knowing at this time what the color of the plant was, but the presence of stomata in the epidermis of the aerial stems (Fig. 4-2B) would seem to be a good indication that cells near the surface of the stem functioned in photosynthesis.

A small conducting rod, or *stele,* runs through the center of the stem. It consists of a cylindrical strand of slender, elongated cells just a few cells thick (see Frontispiece). These cells in the very center, called *tracheids,* are the wood, or *xylem,* and are among the simplest kinds of wood cells known. They resemble cells of the surrounding tissues except for conspicuous rings of thickenings on the inner faces of their thin, presumably cellulose, walls (Fig. 4-2C). Because the thickenings are in the form of rings, these tracheids are referred to as *annular* tracheids. It is assumed that a narrow zone of light cells immediately surrounding the xylem represents the food-conducting part of the stele, or the *phloem* (see Frontispiece). This very simple kind of cylindrical stele with a solid core of xylem is generally considered the most primitive kind of stele by plant anatomists.

In this plant, sporangia that terminate some of the branches are quite unspecialized. In fact, the same epidermis and outer cortex are common to both the stem axis and the sporangium. Within the sporangium are frequently preserved numerous spores, often in groups of four (tetrads). This feature is evidence that the part of the plant known is the sporophyte, and that meiosis occurred in the sporangia of diploid plants.

Rhynia is but one of a number of extremely simple plants that existed in the Silurian and Devonian Periods. Its structure is actually the simplest known of any vascular plant, with the exception of certain modern plants—duckweeds, for example—that are simple as a result of extreme reduction.

Coexisting with plants such as *Rhynia* was another group of simple plants that are preserved in the fossil record. These plants were very similar in that they, too, were composed of dichotomizing axes. They differed, however, in that sporangia were borne along the sides of the axes rather than terminally (Fig. 4-3). Sometimes there were sharp, spinelike projections along the surfaces of the axes as well. Although these projections cannot be considered to be leaves, they did increase the surface area, hence the effective photosynthesizing surface, of these plants. These two groups of early vascular plants—those with terminal sporangia as in *Rhynia* (rhyniophytes) and those with lateral sporangia (zosterophyllophytes)—were quite possibly the ancestral sources of all other groups of vascular plants.

In the same fossil beds that yielded remains of *Rhynia* are other fossilized vascular plants, one of which, *Asteroxylon* (Fig. 4-4), shows a body plan that is still relatively simple but larger and more complex than that of *Rhynia.* From a horizontal, naked, stemlike axis arose aerial branches that were covered with many small, crowded, leaflike structures. These "leaves" apparently were not vascularized, and for that reason botanists hesitate to refer to them as true leaves. From the basal, horizontal axes arose other branches that penetrated the soil. Again, these were not true roots as we know them but, instead, rootlike axes that are morphologically similar to stems but having the same function as roots.

Not only is *Asteroxylon* more complex externally than a plant like *Rhynia,* but the internal structure also shows more elaboration. The stele, though still relatively simple, is much larger than that in *Rhynia;* the xylem is a solid strand that has a conspicuously lobed margin in cross section (Fig. 4-5). These lobes, of course,

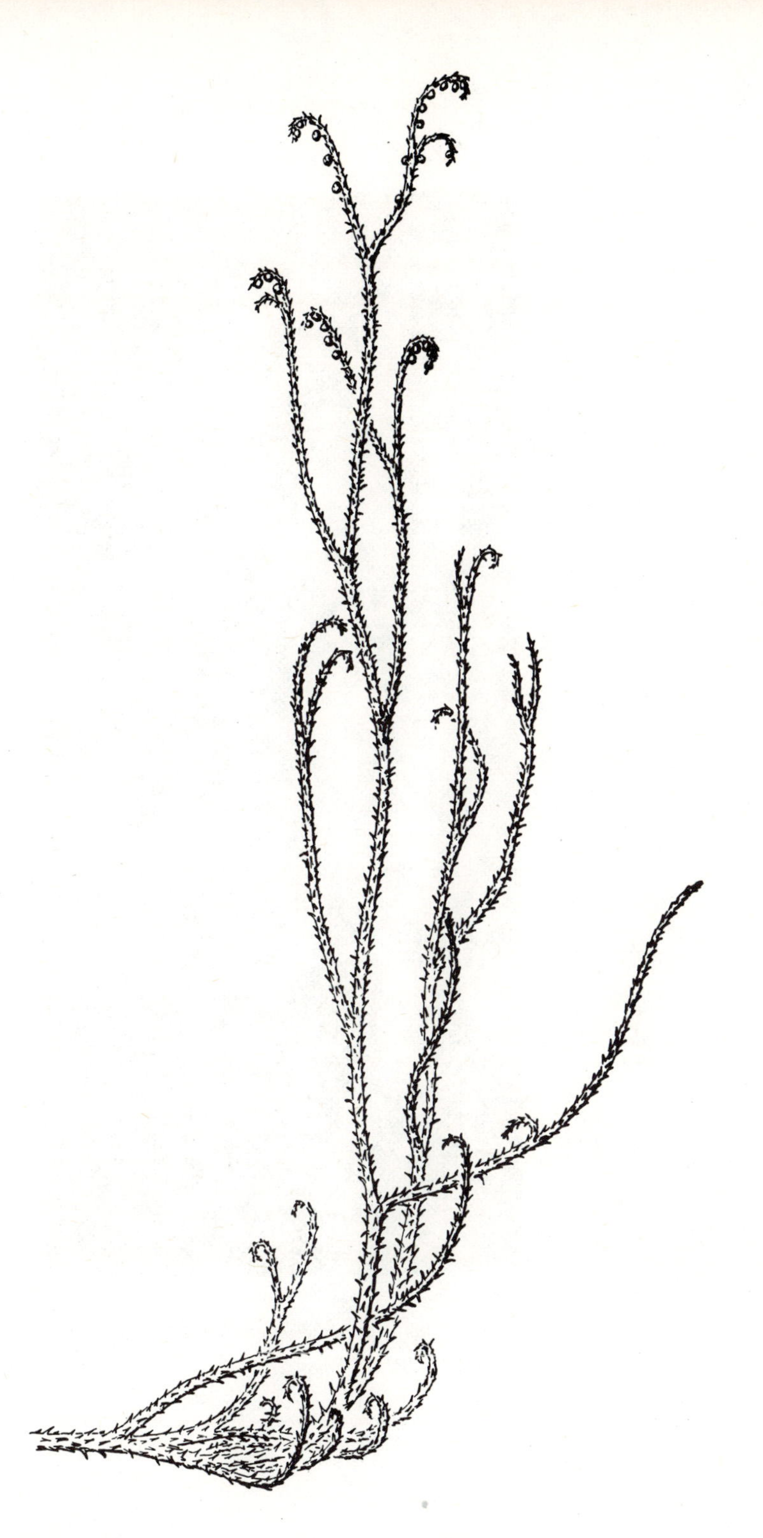

Fig. 4-3 Sawdonia ornata, *a Devonian zosterophyllophyte with lateral sporangia and spiny axes.* ×*0.5.* (*From H. N. Andrews,* Annals of the Missouri Botanical Garden, *vol. 61, p. 186, 1974.*)

Fig. 4-4 Asteroxylon mackiei, *reconstruction.* $\times 0.5$. [*With permission of the trustees of the British Museum (Natural History).*]

are actually ridges running parallel with the axis. Tracheids in the xylem of *Asteroxylon* are also of a more advanced type. On the inner faces of the tracheid walls are secondary thickenings in the form of tightly coiled helices. For that reason these tracheids are often spoken of as *helical* or *spiral* tracheids.

Of significance is the fact that tiny vascular strands originate from the arms of the stele and have an upward and outward course through the cortex in the direction of the leaflike scales. These vascular bundles, however, terminate in the outer cortex and never actually enter the scales. Paleobotanists have interpreted this situation as indicating an early stage of evolution of certain kinds of leaves. In fact the scales of *Asteroxylon* may be considered the equivalents of the spinelike projections mentioned above that occur in some of the early vascular plants with lateral sporangia. Although technically the scales of *Asteroxylon* are not regarded as leaves, functionally they perform as leaves and serve to increase the photosynthetic surface of the plant. The stage represented in *Asteroxylon* could be interpreted as one so early in the evolution of leaves that the vascular bundle had not yet formed the vein of the leaf.

Recent work on the genus reveals that *Asteroxylon* bore small, kidney-shaped sporangia along the stem axes interspersed among some of the scales. The sporangial

Fig. 4-5 Asteroxylon mackiei, *transverse section of the stem with its lobed stele.* × 7.5.

position is good evidence that *Asteroxylon* had as its ancestors simple vascular plants with lateral sporangia (zosterophyllophytes).

It is only a very short step from an *Asteroxylon* type of body plan to that of a typical club moss. *Lycopodium* is a well-known genus of extant club mosses that has a variable habit. In many there is a dichotomously branched stem system that is clothed with small, scalelike leaves. These leaves are actually vascularized, however, so they may be regarded as true leaves. Small, branched roots arise from various parts of the stem and anchor the plant to the ground. These structures, unlike the anchoring devices in *Asteroxylon*, are true roots and not simply branches of the stem. On the upper surfaces of some of the leaves are small, kidney-shaped sporangia (Fig. 4-6B).

Superficially there is little difference between the general habit of the Devonian *Asteroxylon* and a species of *Lycopodium* such as *L. lucidulum* that lives about four hundred million years later (Fig. 4-6B). Even anatomically there is considerable similarity. In some species of *Lycopodium* the stem has a stele with xylem lobed in cross-sectional aspect (Fig. 4-6D). Vascular bundles that supply the leaves arise from the tips of the arms as in *Asteroxylon.* In *Lycopodium,* however, these leaf traces enter the leaves and are the source of the veins in the leaves. Each leaf has a single strand of vascular tissue that extends for its entire length. *Lycopodium* xylem is somewhat more advanced in that the tracheids have considerably more secondary wall thickening material, with the thickenings continuous at the corners, and ladderlike connections along the faces (Fig. 4-6C). Such tracheids with ladderlike secondary thickenings are called *scalariform.*

In a series such as that just described for *Rhynia, Asteroxylon,* and *Lycopodium* (it should be emphasized that this series should not be interpreted as a direct evolutionary line with implied relationships among the members), it is possible to recognize the progressive increase in complexity of the sporophyte from an extremely simple, leafless, rootless plant to one with more elaborate structure, possessing true leaves and roots. All of these changes are most likely reflections of a progressive adaptability to a land environment where the sporophyte generation plays an increasingly important role in both vegetative and reproductive aspects.

Members of the lycopod division, however, had a body plan that did not go much beyond the still relatively simple structure typified by *Lycopodium.* Admittedly, some of the ancient lycopods, such as the giant arborescent forms that existed 350 to 250 million years ago during the Carboniferous and Permian Periods, were more complex than *Lycopodium,* but they deviated only slightly from its basic structural plan.

The evolution of more complex body plans in other groups of vascular plants followed a different course, but these too are well documented in the fossil record. A plant such as *Rhynia* could again be used as the basic starting point. It should be emphasized once more that this does not suggest that *Rhynia* itself was the actual ancestor. It was only one of many kinds of early vascular plants that had the same kind of simple structure, typified by a naked, dichotomously branched, green axial system.

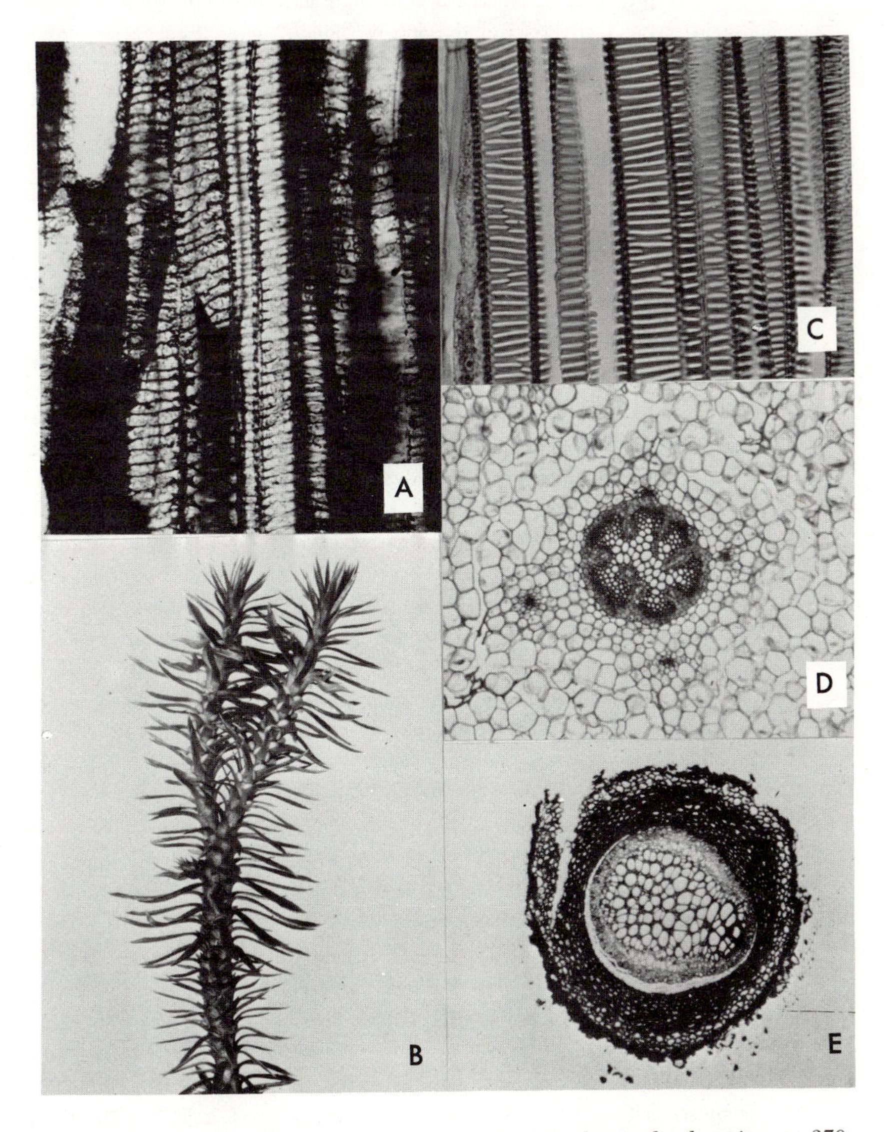

Fig. 4-6 *Helical tracheids of* Asteroxylon mackiei *in longitudinal section.* × *270.* (*B*) Lycopodium lucidulum, *tip of plant. Note sporangia on upper sides of some of the leaves.* × *1.5.* (*C*) *Scalariform tracheids of* Lycopodium inundatum *in longitudinal section.* × *350.* (*D*) Lycopodium lucidulum, *transverse section of lobed xylem in the stele.* × *80.* (*E*) Lygodium palmatum, *cross section of stem, with a solid xylem cylinder in the stele.* × *18.*

There also coexisted during the Devonian Period certain other leafless vascular plants, which, rather than being strictly dichotomously branched, had frequently unequal daughter axes, with one member developing more robustly than the other (Fig. 4-7). This type of branching gave the impression of a more or less "main" axis, with less well-developed "lateral" ones. Such a situation is often referred to as "overtopping"; that is, one member of the branching rises above the other. (These plants are placed in the division *trimerophytes.*)

Progressive Specialization of Leaves

During the middle and latter portion of the Devonian Period there existed a number of fairly large plants with an obviously unequal type of branching and with the lateral appendages considerably divided. Even though this entire lateral system may have represented simply the ramifications of a branching system, the overall appearance of these distal parts was like that of a large fern frond. Obviously the tips must have been photosynthetic, so the whole lateral branch system functioned as a leaf. In some of these fossil plants (for example, the Upper Devonian *Archaeopteris,* shown in Fig. 4-8), a flattening of the very ultimate parts, together with development of photosynthetic tissue between the segments (*webbing*), could have produced wedge-shaped leaves, with the veins reflecting the original dichotomous nature of the ultimate branch segments. Planation of a larger segment of the branching system would result in a large leaf resembling the frond of a fern. In fact, venation of many kinds of fern leaves shows a dichotomous pattern. During the next geologic period (Carboniferous) there were many plants with massive fernlike leaves, the ultimate segments of which closely resemble lateral branching systems of Devonian plants. We speak of leaves formed from such a flattened and modified branch system as *megaphylls* and of plants that bore them as being *megaphyllous.* Fern leaves are excellent examples of megaphylls that were formed phylogenetically by overtopping, planation, and webbing.

Such leaves differ considerably in their origin from those of *Lycopodium,* described previously. Lycopod leaves were probably derived from simple outgrowths of the outer tissues of the stem. Earliest evolutionary stages showed nonvascularized scales; later stages show a vein running through the length of the leaf. Leaves derived in this manner are called *microphylls,* and plants that bore them are called *microphyllous.* Unfortunately the term "microphyll" implies a small size. Certain microphyllous plants, namely Carboniferous and Permian lycopods, actually had extremely long leaves, some of them as long as a meter. The term "microphyll" refers to the origin of the leaf rather than to its size, and it is thought by many botanists that even though certain fossil lycopod leaves may have been long, their origin was the same as that of *Lycopodium* leaves.

Although microphylls and megaphylls may have had different origins, it is clear that they have a similar function and that the end result of the evolutionary series leading to true leaves is relatively similar in both. Both microphyllous and megaphyllous plants had a convergent type of evolution. The body plan is generally

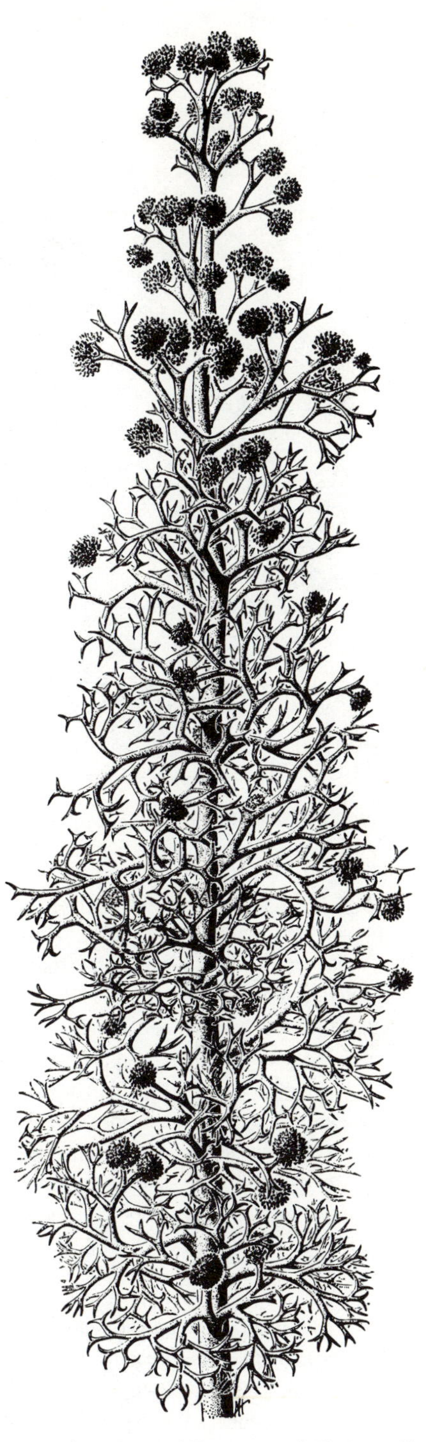

Fig. 4-7 *Reconstruction of a plant of* Pertica quadrifaria, *a Devonian trimerophyte.* ×*0.20. (From A. E. Kasper and H. N. Andrews,* American Journal of Botany, *vol. 29, p. 909, 1972.)*

Fig. 4-8 *A flattened branch of* Archaeopteris, *with wedge-shaped leaves borne on a flattened branching system.* $\times 0.45$.

the same, with a stem system bearing appendages, or leaves, of a different nature; in both kinds of plants the photosynthetic area is increased considerably. Increase in food-manufacturing potential obviously allows for greater size of the plant. Greater size of the sporophyte, as will be seen in a later section, can permit an increase in reproductive potential.

All megaphylls are not fernlike in appearance. In fact, most leaves in the plant kingdom are not divided and frondlike but are entire, uninterrupted units. The evolution of such kinds of leaves is interpreted as the result of further "webbing" of the divided leaves, with a union of the individual parts into a single blade or lamina. The veins of entire or undivided leaves are interpreted as the vestiges of the dichotomizing axes of ancient ancestors. Subsequent evolution has resulted in a fusion of some of these axes to produce an anastomosing network of veins.

Other Specializations of the Sporophyte

Along with progressive specialization of leaves in vascular plants occurred other specializations of the sporophyte. With continued increase in photosynthetic area and more elaborate vascularization, the sporophyte generation became larger. Some forms assumed treelike stature even as early as the Devonian Period. A clustering of spore-producing members occurred along with this increasing complexity of the sporophyte. In the lycopods, for example, the sporangium-bearing leaves became aggregated into cones (Fig. 4-9). A similar arrangement of sporangium-bearing branches of the horsetails or articulates occurred. In the ferns, sporangia tend to be grouped into little clusters on the undersides of the fronds. Seed plants, too, tend to produce spore-producing members in aggregations.

Although the tree habit is obviously derived from a more dwarfish, nearly herbaceous habit, many relatively primitive groups of plants remained herbaceous. *Lycopodium,* for example, differs little from ancestors in the Paleozoic Era. However, some of the lycopods were huge trees during the Carboniferous and Permian Periods, having become considerably more elaborate than the earlier lycopod progenitors. Many ferns as well apparently have a persistently primitive habit, never having attained tree proportions. Others were once arborescent.

Fig. 4-9 Lycopodium obscurum, *the tip of a plant with sporangium-bearing leaves aggregated into cones.* $\times$ *1.2.*

In contrast to the persistent herbaceous types of vascular plants are the angiosperm, or flowering plant, herbs. These plants, which form so much of the green cover of the earth, are relatively recently evolved, and it appears that they may have originated as a response to the progressive cooling of the earth's surface during late Mesozoic and Tertiary times. Full implications of the origin of herbaceous angiosperms are discussed in Chapter 5.

EVOLUTION OF THE STELE

Intimately related to the increasing complexity of the sporophyte phase of vascular plants is the elaboration of the vascular system. The conducting cylinder, or stele, of sporophytes began first as a simple solid strand only a few cells in thickness (for example, *Rhynia*). Not especially evident in many of the fossils, but quite evident among living plants, is a zone of tissue called *phloem* that surrounds the xylem core. Whereas the principal function of the xylem is water and mineral conduction, phloem functions primarily in transporting dissolved foods, particularly carbohydrates. These primitive, solid steles are found not only in ancient plants like *Rhynia* but also in some lycopod axes, in some ferns (Fig. 4-6E), in many kinds of roots, and in other places as well.

A solid stele with a xylem margin that is star-shaped, or lobed, in cross section is a somewhat more advanced kind of stele, but is found also in many primitive kinds of plants (for example, *Asteroxylon*, certain lycopods, and many roots; see Figs. 4-5, 4-6D). Very often in these lobed steles the phloem occupies the position between the xylem lobes, in the furrows.

In these simple steles the principal water- and mineral-conducting elements of the steles are extremely slender, elongated cells, with tapered ends. After the secondary wall thickenings are deposited on the inner faces of the primary walls, the cells die and function as water- and mineral-conducting cells in a nonliving state. Certain botanists believe that these nonliving tracheids must be in close association with living cells of some kind. Although there is no known physiological basis for this assumption, these botanists believe that because of the very definite close association between nonliving conducting cells and living cells, there must exist some unexplained relationship between them.

While a solid stele is small, each tracheid is not very far from a living cell surrounding the xylem. If the size of the solid stele were increased with the addition of more living cells, however, many tracheids would be a considerable distance from living cells. Yet by formation of ridges and furrows, as in the *Asteroxylon* stele, or by an interspersion of living cells among the tracheids there can be a considerable increase in the actual size of the stele, and many more tracheids can be present without having any given tracheid too far from a living cell. Modification of the stele to allow such an association of nonliving and living cells is often called the "principle of size and form." This principle implies that the form or shape of a stele is influenced by the size it attains. It would imply, for example, that a solid stele would never get to be extremely large if the xylem consisted only of tracheids (and, indeed, none has ever been found); with increase in size, the stele undergoes certain modifications that maintain the ratio between living and nonliving cells.

Formation of ridges and furrows is only one example of stelar modification accompanying increase in size of the stele. Certain plants are known to have steles in which lobing has become so pronounced that the sinuses of the xylem often extend completely through the xylem cylinder (Fig. 4-10A). In transverse section the stele appears to be composed of a series of platelike segments of xylem with phloem between. Again, it is obvious that no tracheids are at any appreciable distance from a living cell.

Other kinds of vascular plants have means other than formation of ridges and furrows to maintain the nonliving-living cell relationship in the stele. In some, the stele became large (phylogenetically speaking), and the center of the stele remained living, never having become modified into tracheids. This central tissue is called a *pith,* and is found in most of the higher vascular plants. Steles with pith in the center, surrounded by xylem, which in turn is surrounded externally by phloem, are commonly found in some lycopods and in many ferns and seed plants. In certain vascular plants, ferns especially, there is a zone of phloem both on the outside and on the inside of the cylinder of xylem (Fig. 4-10B).

Effect of Leaves upon the Stele

The kinds of steles cited above are the principal kinds of conducting cylinders in vascular plants, but there are many kinds of modifications of them. These modifications are due primarily to the effect leaves have upon the stele.

In plants with solid steles, as well as in lycopods with steles containing pith, strands of conducting tissue (vascular bundles) that originate from the stele and extend upward and outward through the cortex into leaf bases have no effect on the stele. These bundles, or *leaf traces,* simply "pinch" off from the stele and really involve a very small part of the stele.

In other plants, especially ferns and seed plants, a leaf trace is actually composed of a considerable part of the stele and the xylem and phloem are interrupted above the level of its separation from the stele (Fig. 4-10C). This interruption is "closed" at a higher level where the stele once again becomes a continuous cylinder. This break in the stelar cylinder is called a *leaf gap,* but it must be emphasized that the term "gap" is not synonomous with "space." The region of the leaf gap is actually filled with tissues, generally thin-walled cells that are continuous from the pith to the cortex surrounding the stele.

If leaves are distantly spaced on a stem, it is obvious that leaf gaps will be distantly spaced; seldom will more than one gap be visible in any cross section. In a plant with numerous, closely spaced leaves, however, there may often be several gaps at a given level. Thus when it is viewed in cross section, the stele appears to be composed of a number of separate vascular bundles arranged in a cylinder (Fig. 4-10D; in this plant only the xylem is dissected, with the phloem uninter-

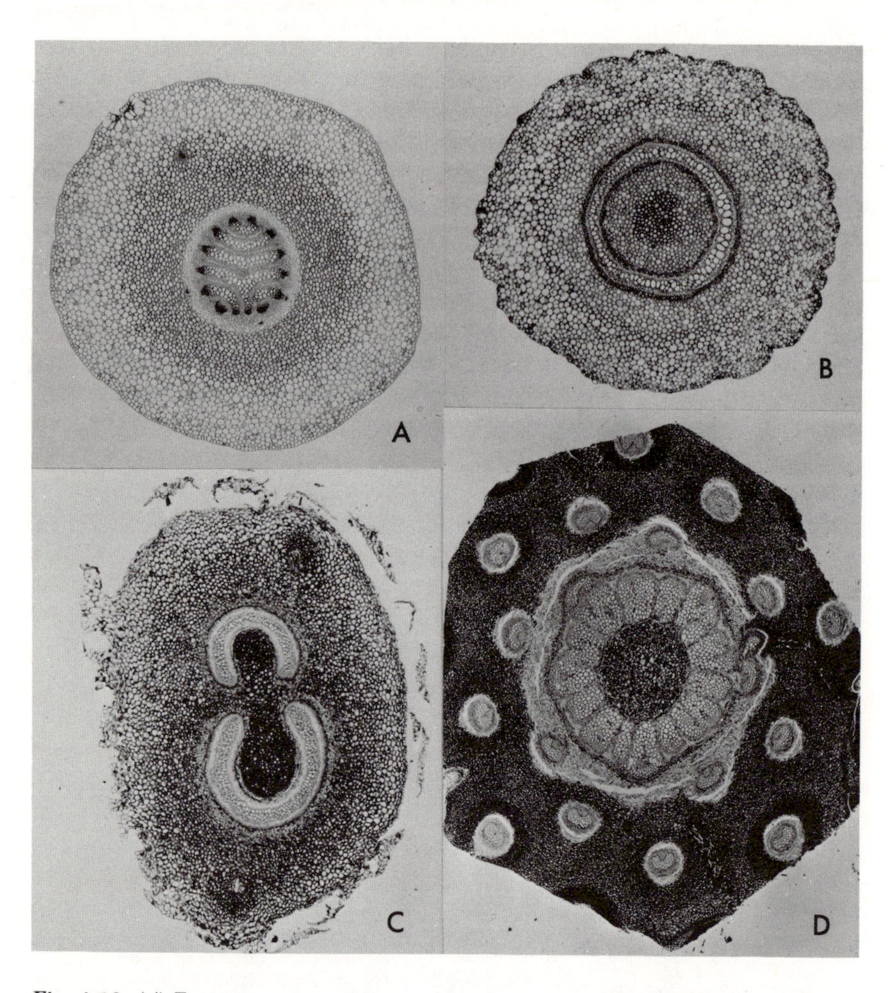

Fig. 4-10 (*A*) *Transverse section of the stem of* Lycopodium complanatum. ×*18.* (*B*) *Stem of hay-scented fern,* Dennstaedtia punctilobula, *in transverse section.* ×*17.* (*C*) *Maidenhair fern,* Adiantum pedatum, *stem in cross section. Note gap left in stele above the level of separation of a leaf trace* (*upper part of picture*). ×*11.5.* (*D*) Leptopteris superba *stem in transverse section. The xylem is dissected by many gaps.* ×*6.7.*

rupted). It should be remembered, however, that these gaps are "closed" at higher levels, and that the stele is basically a cylinder with gaps in it resulting from the separation of leaf traces. Among the seed plants are steles that are considerably dissected, not only by leaf gaps but by other interruptions as well.

Flowering plants of the monocotyledonous type differ considerably from the forms of plants having the stelar types just described. In stems of these plants (such as grasses, lilies, palms, and orchids) many bundles are scattered throughout a soft tissue with no recognizable pith or cortex (Fig. 4-11A).

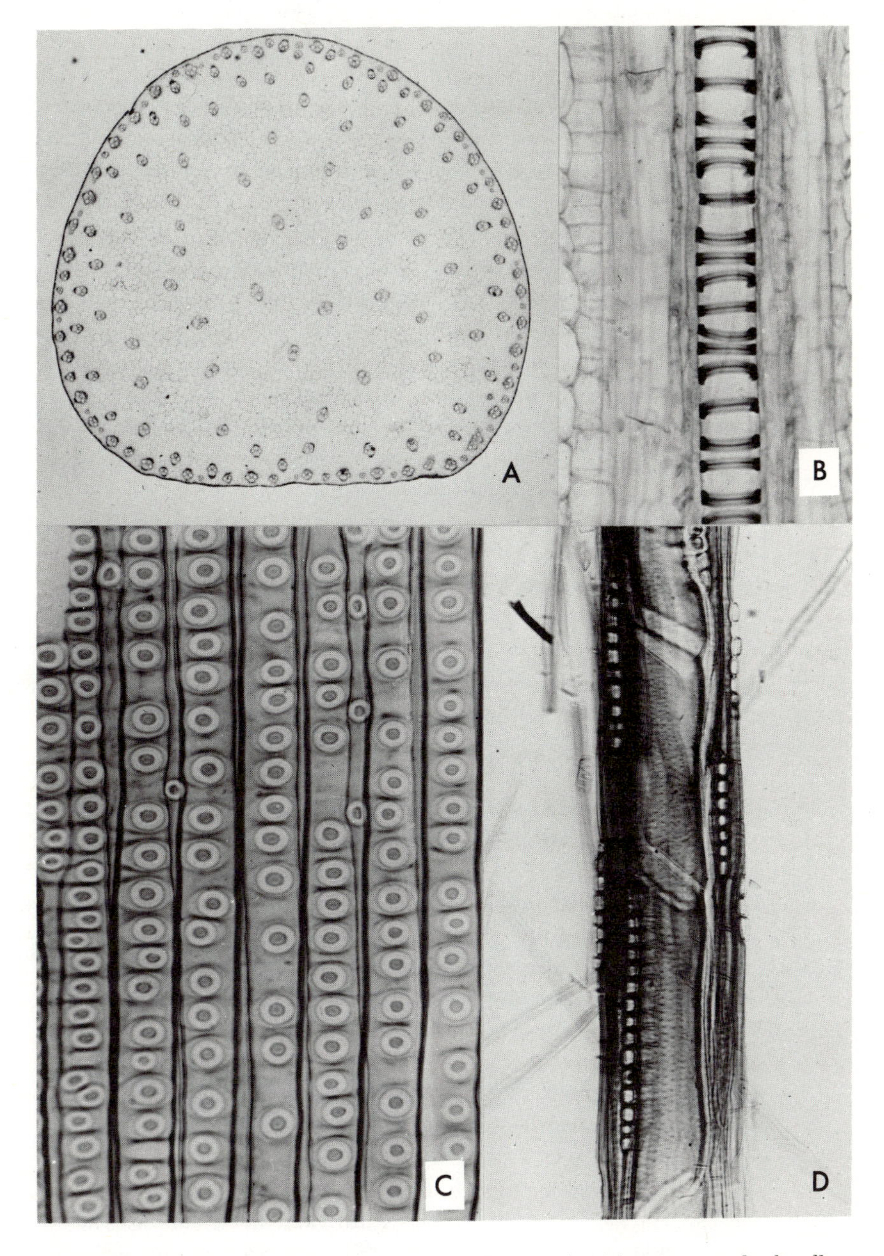

Fig. 4-11 (*A*) *Transverse section of a corn stem* (Zea mays), *with many vascular bundles visible.* ×*8.5.* (*B*) Annular xylem elements of a corn stem. ×*260.* (*C*) *Circular bordered pits in face view in the secondary wood of white pine* (Pinus strobus). ×*240.* (*D*) *Vessel in oak wood* (Quercus coccinea). *Chemical treatment has resulted in the partial separation of the individual vessel elements.* ×*155.*

The kinds of steles described form the basic skeletal network of conducting tissue. Cells of the conducting tissue are derived simply by a maturation of new cells that are produced by the growing tip of the shoot or the root. Xylem and phloem formed by maturation of cells produced at the apex are called *primary xylem* and *primary phloem.* Steles composed only of primary tissues seldom get large or massive. The bulk of wood and phloem in large vascular plants, such as stout herbs, shrubs, and trees, is produced later, after maturation of the primary xylem and phloem. (There are a few notable exceptions of plants that attain a large size with only primary tissues—tree ferns and palms, for example.) In these plants there is to be found between the primary xylem and primary phloem a thin band of thin-walled cells with the potential of dividing, called *vascular cambium.* When this cambium divides, it produces more xylem cells on the inner side and more phloem cells on the outside. The bulk of a tree therefore is xylem produced by the vascular cambium; phloem is produced in lesser amounts at the outside of the cambium. Xylem and phloem produced by the vascular cambium, called *secondary xylem* and *secondary phloem,* often mask the original stelar structure represented by the primary xylem and phloem.

Development of secondary xylem and phloem is a feature of many of the vascular plant groups. It is generally absent in the rhyniophytes, present in the lycopods (especially among extinct Paleozoic forms), present in the horsetails (only in extinct Paleozoic forms), rare in the ferns (although occasionally present in Paleozoic fossils), and quite common in the seed plants, both living and fossil.

EVOLUTION OF XYLEM ELEMENTS

The evolution of the conducting cells in the xylem seems to be related to the developmental stages of the various groups of vascular plants. Annular (Fig. 4-11B), helical (Fig. 4-6A), and scalariform (Fig. 4-6C) elements have been mentioned already. Annular tracheids are found as the sole kind of xylem-conducting element only among the most primitive vascular plants, such as some rhyniophytes. Helical tracheids are the sole kind of xylem-conducting cells in only a few plants, namely rhyniophytes and zosterophyllophytes. Scalariform tracheids are somewhat more specialized and are found frequently in the lycopods and in the ferns. All three types, however, may be found in one plant, with the first xylem cells to mature being annular, somewhat later ones helical, and the last formed ones scalariform.

A still more advanced kind of tracheid is one with considerable secondary wall thickening on the inner surface of the primary wall. In certain places this secondary wall is raised into dome-shaped structures projecting into the cell cavity; each dome-shaped structure has a small pore at the top. These structures occur in pairs—that is, they will be developed by two adjacent cells at corresponding spots. These modifications of the wall are called *bordered pits* (Fig. 4-11C). Bordered-pitted xylem cells are the most common in the seed plants and in the horsetails; they also occur in a very few ferns.

Even in plants with circular bordered pits as the characteristic type there may be evidences of the more primitive kinds of tracheids. Frequently the first xylem cells to mature in a stem do so before the region in which they are found has ceased to elongate. The thin cellulose primary walls are capable of stretching, and the secondary thickenings in the form of strengthening rings tend to keep the cell walls from collapsing completely. Somewhat later, helical cells may mature, and these also may be stretched, with the helical thickenings simply opened out. Only when elongation in a given region has ceased are scalariform and circular bordered-pitted cells formed.

While the conducting cells of the xylem in the so-called lower vascular plants are typically tracheids, the angiosperms, or flowering plants, have evolved a much more efficient water- and mineral-conducting mechanism. In the wood of most angiosperms are structures called *vessels,* composed of a series of tubular structures that are arranged end to end, with the wall dissolved away at the places where adjacent cells touch (Fig. 4-11D). Vessels, therefore, are not individual cells, but tubular, canallike structures made up of a series of cells. In the most advanced kinds of vessels the individual vessel elements are short and squat, with large circular pores at both ends of the cells. These vessel elements are aligned in such a way as to resemble a series of drain tiles and form a continuous tube.

Although vessels are found primarily among the angiosperms, a few gymnosperms (naked-seeded plants) and ferns have them. *Selaginella,* one of the lycopods, has them, and *Equisetum,* a horsetail, sometimes has vessels.

In spite of the occurrence of vessels in vascular plant groups other than the angiosperms, it is with the latter group that vessels are most often associated. Their almost universal occurrence in the flowering plants is another example of the many features in the angiosperms that are more highly specialized than those in other groups of vascular plants. Certain angiosperms, however, lack vessels. Some of these are woody plants in families considered to be primitive among the flowering plants. In these plants it seems most likely that vessels had never evolved and that the plants are in groups that were always vesselless. In other flowering plants that lack vessels it seems most likely that their absence is a result of a loss and that ancestors of these forms did possess vessels. Certain aquatic plants, for example, as well as some cacti represent forms in which the absence of vessels may be a secondary development.

EVOLUTION OF VASCULAR PLANT REPRODUCTIVE SYSTEMS

Discussion of vascular plant specialization thus far has been concerned primarily with vegetative aspects, with almost no mention of reproductive systems. Among the vascular plants there is a wide range of reproductive systems, and these systems often serve as bases for classification of the groups. In spite of the apparent diversity of reproductive methods, however, there is actually a surprising degree of uniformity. The differences are merely modifications of a basic underlying theme, although per se they are extremely interesting and represent varying degrees of evolutionary specialization.

The common type of life cycle of all vascular plants will be presented in a general fashion here, and the departures in specific groups will be interpreted as variations of this basic plan. All vascular plants have a sporophyte phase; this is the diploid phase and it is the part of the life cycle in which vascular tissue is developed. It is the dominant part of the life cycle and the most conspicuous part as well. Somewhere on the sporophyte are produced sporangia; the actual position of sporangia varies among the groups, but they are always to be found on the sporophyte. Within the sporangia are cells called spore mother cells, or *sporocytes.* These sporocytes are also diploid, and the number within a sporangium may vary considerably among the various plant groups. Meiosis, or reduction division, occurs within the sporangium when the sporocytes divide. As a result of meiosis spores are produced, and these initiate the haploid phase of the life cycle. Germination of the spores results in the formation of a multicellular body, the gametophyte, which produces gametes. Two kinds of gametes are produced by the gametophytes: small sperm cells (the male gametes) and larger egg cells (female gametes). Fertilization occurs when a sperm cell unites with an egg to produce a diploid zygote—the start of the new sporophyte phase. The whole cycle is repeated again, with the haploid and diploid phases alternating regularly.

This life cycle is the normal one, and certain "abnormal" ones may be dismissed after a brief mention here. For example, in certain ferns, gametophytes may bud off from the sporophyte without spores being produced. Such a phenomenon is termed *apospory.* Similarly, sporophytes may bud off directly from the gametophytes without sexual union. This phenomenon is called *apogamy.* Generally speaking, in cases of apogamy or apospory, the new phase produced by budding has the same chromosome condition as does the phase from which it buds. The normal alternation of chromosome number is missing. Apogamy and apospory, while having interesting cytological and morphogenetic aspects, are not involved in the evolutionary considerations presented here.

Homosporous Plants In what we may consider the most primitive situation in the vascular plants, the sporophyte and gametophyte are both independent plants, each capable of absorbing materials from the substrate and manufacturing its own carbohydrates. A fern plant is a good example of such an arrangement. In the primitive condition all spores produced by the sporangia of a sporophyte are of one type; such a plant is called *homosporous.* Most ferns are examples of homosporous plants. Not only are the spores alike in size and structure, but upon germinating, all spores of a given plant produce the same kind of gametophyte. This gametophyte bears both male and female sex organs (Fig. 4-12A). In these primitive vascular plants, the sperm are produced within a specialized structure called an *antheridium* (Fig. 4-12B). Generally many sperm cells are produced within one antheridium. The egg cell is generally borne singly within a specialized structure called an *archegonium* (Fig. 4-12C). Most fern gametophytes are of this type; that is, the same gametophyte produces both antheridia and archegonia.

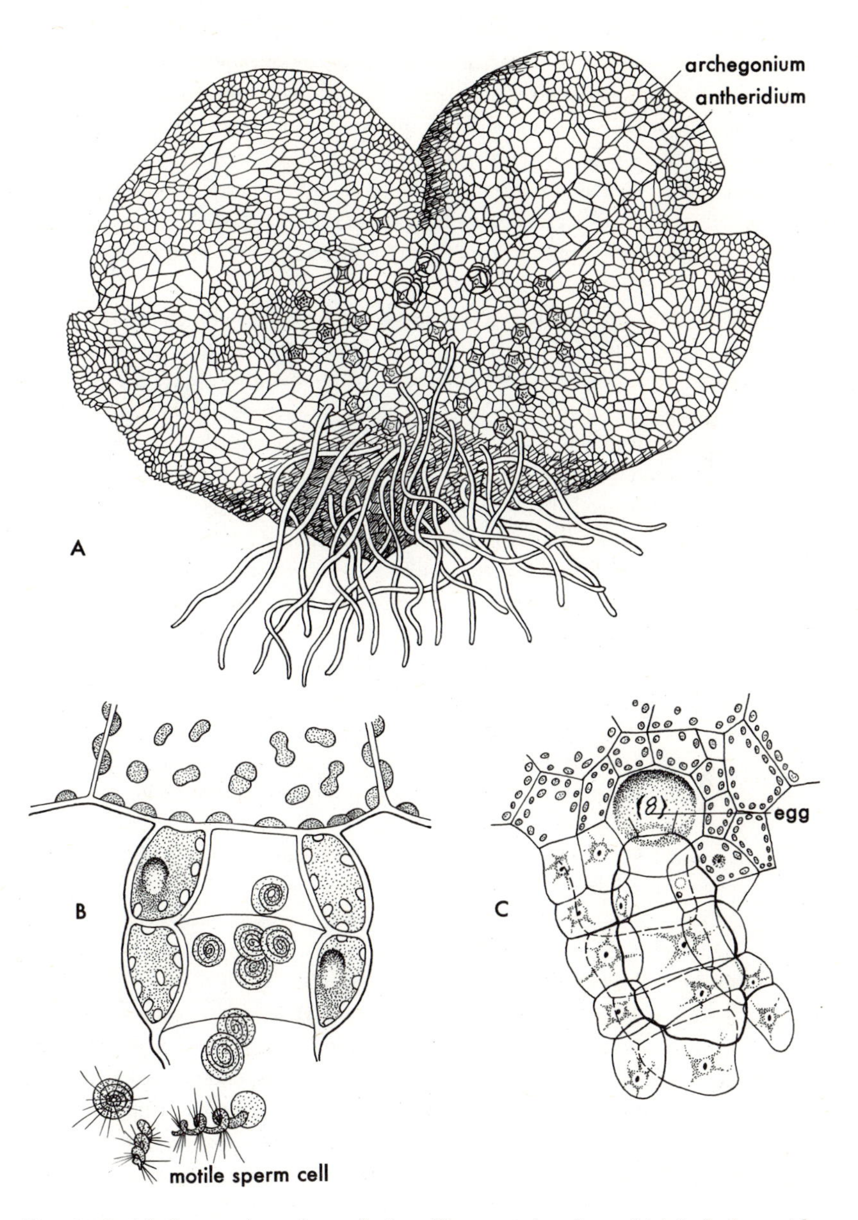

Fig. 4-12 *(A) Gametophyte phase of a fern. The exposed surface, which is the lower side, bears hairlike absorbing structures, spherical antheridia, and archegonia, the necks of which project from the surface. (B) Antheridium and motile sperm cells. (C) Archegonium and egg. [Redrawn from* The Plant Kingdom, *by William H. Brown, © Copyright, 1935, by William H. Brown. Used by permission of the publisher, Ginn and Company, (Xerox Corporation.)]*

When they are mature, the sperm cells are released, and in primitive vascular plants (again, ferns may be used as examples) these cells are motile, having flagella attached to them. The sperm cells swim to the archegonia, and one sperm cell fertilizes one egg. Because the sperm is dependent upon water to reach the egg in the archegonium, a severe limitation of habitat is imposed on the plant. Although the plant is strictly speaking a "land" plant, fern plants are most successful in habitats where there is generally moisture available in the form of rainwater, dewdrops, or simply accumulated moisture on the ground. For that reason also, the gametophyte generation is never very large. It generally grows close to the ground, and sperm cells are released not far from the ground (or whatever other substrate on which the gametophyte may be growing).

Although fern plants have been used here as examples of homospory, it should be mentioned that the rhyniophytes, zosterophyllophytes, certain lycopods, and living horsetails have a similar habit. In all of these, spores are of one type, and gametophytes, where known, have both kinds of sex organs.

Environmental conditions sometimes influence the production of sex organs on gametophytes, and under certain conditions only antheridia or only archegonia may be produced. In such situations one might confuse homosporous plants with types having two kinds of spores, each producing a different kind of gametophyte.

In a number of plants, notably among the horsetails, all of the spores are alike morphologically, but there are instances in which a gametophyte does not bear both kinds of gametes simultaneously.

Heterosporous Plants A still more advanced condition is one in which sporophytes produce two kinds of spores in two kinds of sporangia. Certain sporangia produce a smaller number of large spores (called *megaspores*), and others produce a greater quantity of smaller spores (called *microspores*). In such heterosporous plants each of the two kinds of spores germinates into a different kind of gametophyte phase. The small spores produce the male gametophyte (or microgametophyte), which produces only sperm cells. Megaspores produce larger, female gametophytes (or megagametophytes), on which only female sex cells, the eggs, are produced. *Selaginella,* one of the lycopods, is an example of such a heterosporous plant (Fig. 4-13A). In this generally herbaceous plant (as in all lycopods) the sporangia, borne on the upper sides of specialized leaves, are aggregated into cones, with certain of the sporangia producing many small microspores and others a smaller number of megaspores (Fig. 4-13B, C). Quite often there is only one functional megasporocyte in a megasporangium, so only four megaspores are produced after meiotic division. The number fluctuates, but seldom are there more than eight megaspores per sporangium.

The male gametophyte is extremely reduced in *Selaginella;* in fact, when the microspore germinates, all of the cells produced are retained for a time within the microspore wall. The microgametophyte is actually nothing but a single antheridium, with a layer of jacket cells surrounding the sperm cells (Fig. 4-14B). Only the

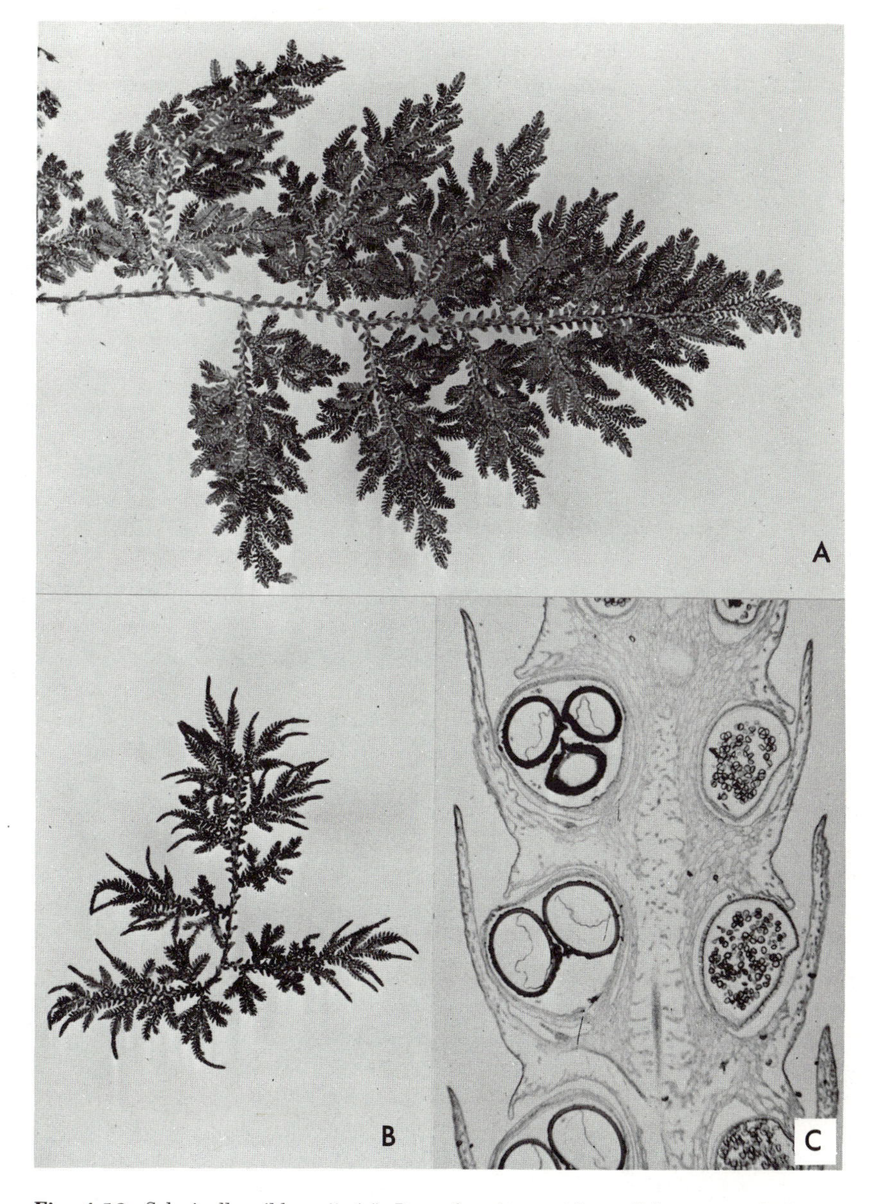

Fig. 4-13 Selaginella wildenovii. (*A*) *Part of a plant, with small leaves attached to a branching axis.* $\times 0.6$. (*B*) *Branches with cones borne at the tips.* $\times 0.6$. (*C*) Selaginella *species, longitudinal section of part of a cone. Megasporangia with enclosed megaspores are on the left side; microsporangia with microspores are on the right.* $\times 40$.

reproductive aspect of the male gametophyte has been retained; the vegetative portion has been completely dispensed with. At maturity the sperm cells are released, usually on or near the ground where the microspores have fallen (Fig. 4-14C).

The female gametophyte of *Selaginella* is also considerably reduced. It is a multicellular haploid body that is retained within the wall of the megaspore during much of its development, but eventually it ruptures the megaspore coat, and part of the megagametophyte may protrude through the break (Fig. 4-14A). Archegonia are borne in the part of the gametophyte that extends beyond the break of the spore wall.

As in the ferns, the sperm cells swim through water and enter the archegonia, where the eggs are fertilized. Again, it is obvious that there is dependence upon water for fertilization, and *Selaginella* plants have limits imposed on the habitats in which they may grow. In order for sexual reproduction to occur, water must be present in some stage of the life cycle. Actually, some species of *Selaginella* may survive in dry and rigorous habitats; they curl up when there has been no rain, but even a small shower is enough to cause them to expand.

Selaginella shows considerable specialization in its life cycle, even though basically it follows the same pattern as do all other vascular plants. Because the sporophyte is the phase of the life cycle best adapted for survival in a land environment, the gametophyte phase is gradually being reduced to its barest

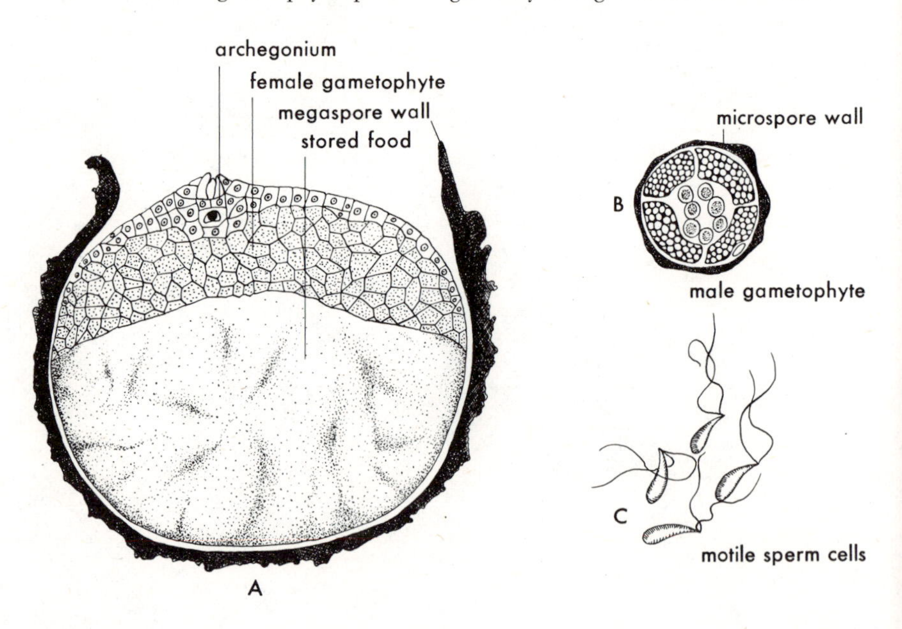

Fig. 4-14 (*A*) *Sectioned female gametophyte of* Selaginella *within the megaspore wall.* (*B*) *Sectioned reduced microgametophte of* Selaginella *within the microspore wall.* (*C*) *Motile sperm cells of* Selaginella *released from the male gametophyte.* [*Redrawn from* The Plant Kingdom, *by William H. Brown,* © *Copyright, 1935, by William H. Brown. Used by permission of the publisher, Ginn and Company* (*Xerox Corporation.*)]

essentials—those parts concerned most directly with production of gametes. Furthermore, food reserves that originated from the parent sporophyte are stored in the megaspores, and this food is utilized by the embryo sporophyte that develops from the fertilized egg in an archegonium. The vegetative, photosynthetic phase has been done away with. In summary, *Selaginella* is representative of a number of vascular plants in which there is heterospory and in which the gametophyte phase has been extremely reduced.

The tendency to change from homospory to heterospory, the reduction in number of megaspores in a sporangium, and finally the extreme reduction of the gametophyte generation were further carried on in a fossil genus of arborescent lycopods. The genus name *Lepidocarpon* is assigned to unusual seedlike bodies that are assumed to have been borne on trees of the type generally called *Lepidodendron,* an important plant during Carboniferous and Permian times (Fig. 4-15). (Because in paleobotanical work it is frequently difficult to demonstrate attachment of various isolated organs, individual names are retained for each of the organs until such time

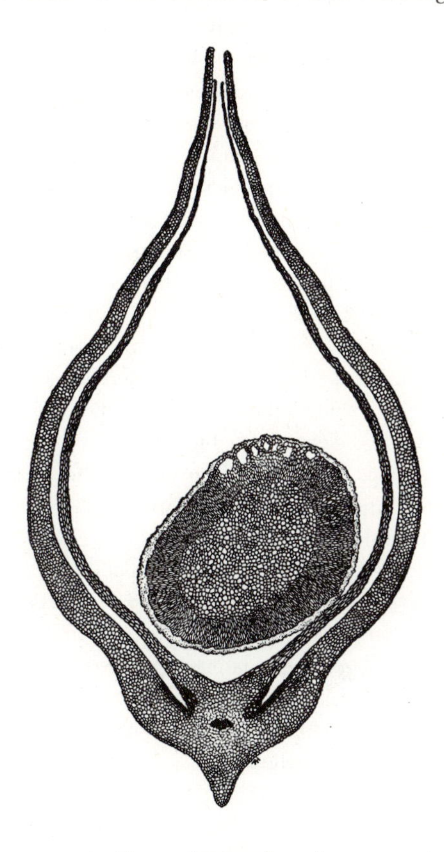

Fig. 4-15 Lepidocarpon magnificum. *Section through a sporangium, with surrounding sporophyll margins and contained megaspore wall with a female gametophyte. (From H. N. Andrews and E. Pannell,* Annals of the Missouri Botanical Garden, *vol. 29, p. 23, 1942.)*

as organic connection can be proved.) As in all lycopods, *Lepidocarpon* sporangia were borne on the upper sides of sporophylls. In this genus, too, sporophylls were aggregated into cones, and both microspores and megaspores were produced. Microsporangia were large, each producing hundreds, or perhaps thousands, of small spores. Megasporangia were also large, but each sporangium produced only one functional megasporocyte. As a result of meiosis four megaspores were produced, but three of them aborted. The single functional megaspore then proceeded to enlarge considerably. It is actually possible to demonstrate that meiosis had occurred in the sporangium and that only one megaspore of four remained functional, for in a megasporangium one can frequently see the three shriveled, aborted megaspores next to the large functional one.

A further specialization in *Lepidocarpon* occurred in the sporophyll, the leaf on which the sporangium rests. The lateral edges of the sporophyll had become considerably extended, covering over the sporangium from either side but leaving a long, narrow slit at the top. What is present is a sporangium, with one functional megaspore that is almost completely enveloped by an integumentary system, in this case the lateral edges of the sporophyll.

Also known from the fossil record is that the gametophyte developed within the megaspore and produced archegonia. Not known, but assumed from what we know of present-day plants, is that the sperm were produced within the microspores in an extremely reduced microgametophyte (much like that in *Selaginella*) and were released at maturity, when they swam into the megasporangia and fertilized the eggs in archegonia that were exposed by a rupture of the megaspore. This would be easily possible because the individual sporophylls of *Lepidocarpon* with megasporangia apparently fell off the tree as units, and fertilization could have occurred on the ground or in the swamps in which these trees were supposed to have grown. The entire megasporophyll unit of *Lepidocarpon* therefore functioned as a seed. An embryo was presumably produced in the megagametophyte within a mass of food-storing tissue, the entire structure being enveloped by what is analogous with a seed coat (the flaps of sporophyll enveloping the sporangium).

In spite of this extreme specialization in *Lepidocarpon* (and in certain other Paleozoic plants), the basic vascular plant life cycle is still discernible. A sporophyte (the large tree) produced on itself many sporangia (the *Lepidocarpon* micro- and megasporangia) that produced spores after reduction division (the many microspores in the microsporangia and the one functional and three aborted spores in each megasporangium). Upon germinating, the spores produced a new phase of the plant, the gametophyte (extremely reduced in the microgametophyte and retained in the spore in the megagametophyte). Sperm cells fertilized the eggs, and the zygotes initiated the new sporophyte phases.

Reproductive Specialization in Seed Plants

True seed plants may be used to continue the story of the extreme specialization in the reproductive cycle. *Ginkgo*, the maidenhair tree (division Ginkgophyta), serves as an example of a fairly primitive situation among seed plants, and pine (conifer division) as an example of a later evolutionary stage.

The *Ginkgo* tree that we see cultivated in parks and gardens or as a street tree represents the diploid sporophyte stage. It is a woody plant with peculiar fan-shaped leaves that have dichotomously branched veins (Fig. 4-16). Most of the leaves are borne on dwarf branches that arise from elongated branch axes. These plants are deciduous, and a *Ginkgo* tree in the autumn and winter is devoid of leaves. In the springtime new leaves are formed at the tips of the dwarf shoots, and in some trees, those that are strictly pollen-producing, slender pollen-bearing cones are produced (Fig. 4-16A). These are elongated axes that bear helically arranged appendages, each with a pair of pendant microsporangia or pollen sacs. Numerous microsporocytes are

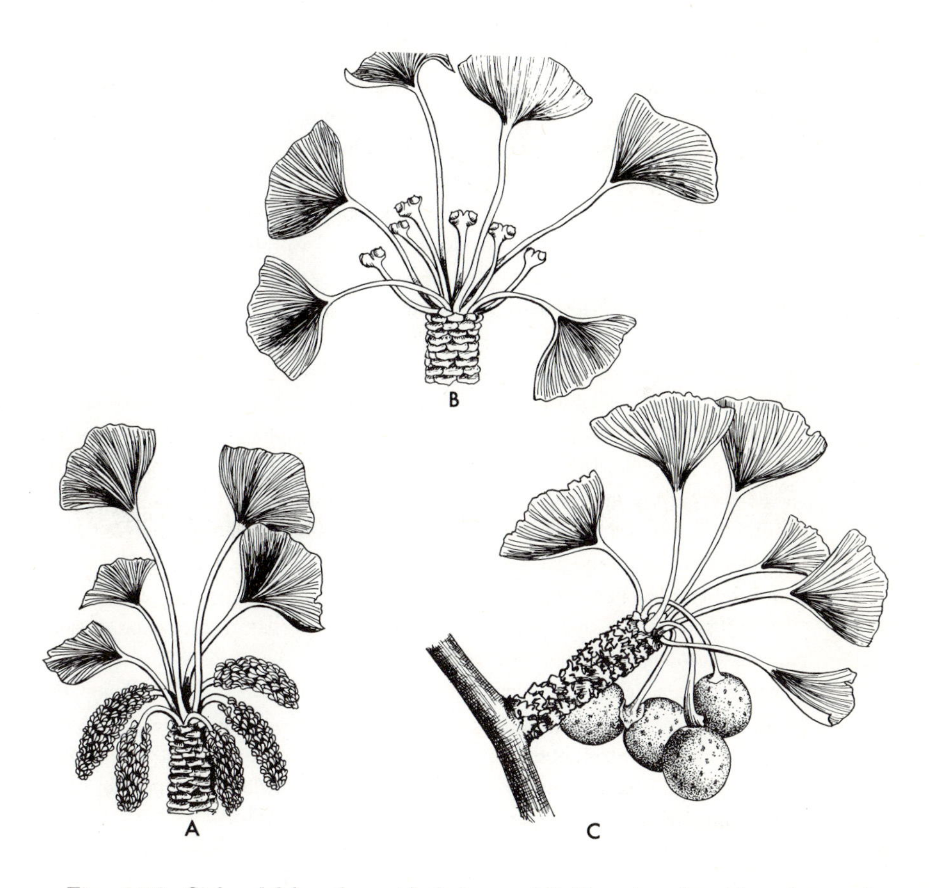

Fig. 4-16 Ginkgo biloba, *the maidenhair tree. (A) Short branch, with a cluster of fan-shaped leaves and pollen-bearing cones. (B) Dwarf branch, with young ovules. (C) Branch, with mature seeds. [Redrawn from* The Plant Kingdom, *by William H. Brown, © Copyright, 1935, by William H. Brown. Used by permission of the publisher, Ginn and Company (Xerox Corporation.)]*

produced in each and, after meiosis, many haploid microspores. These spores do not remain uninucleate before being shed, however, and a mitotic division produces two cells within each. At this stage they are called *pollen grains,* which actually represent an early stage in microgametophyte development. They are shed from the split microsporangia in the binucleate condition.

The seed-bearing axes are produced on other trees. These axes are also borne among leaves on the dwarf shoots; an ovule-bearing stalk is an elongated structure with two young ovules at the tip (Fig. 4-16B). Each ovule is actually a megasporangium encased within an integumentary system, a modification of the sporophyte that bears the sporangium. It would be incorrect to say that this integument is the structural equivalent of the so-called integument of *Lepidocarpon,* but both seem to have the same function. More on the possible origin of integuments of seed plants will be presented later in this chapter. Each sporangium (here there is a change in terminology, and a megasporangium in a seed plant is called a *nucellus*) has but one megasporocyte (Fig. 4-17), and after meiosis four haploid spores are produced in a linear arrangement. The three toward the distal end of the sporangium abort, leaving only one functional megaspore (again paralleling *Lepidocarpon*). This single megaspore then germinates to produce within itself the female gametophyte. The entire ovule enlarges, and eventually within the megaspore membrane is produced a bulky gametophyte with a considerable amount of stored food. At the distal end several elongated archegonia develop, embedded within the gametophyte. In each there is a large egg. Before this stage the integument has almost completely enveloped the nucellus (megasporangium), leaving only a minute tunnellike hole, the *micropyle.*

In *Ginkgo* and the other seed plants the sperm are not released to make their independent way to the eggs. Instead the entire pollen grain is carried from the pollen-producing tree by air currents to an ovule-producing one, and a few land at the tip of an ovule. In many gymnospermous seed plants a mucilaginous droplet is exuded from the micropyle to which pollen grains adhere. As the droplet dries, it pulls the pollen grains into the micropyle against the nucellus. It is here that further development of the microgametophyte occurs. The pollen grain wall ruptures, and the microgametophyte develops into a tubelike structure that grows into the nucellus, toward the megagametophyte. The number of cells in the microgametophyte increases to six, including two sperm cells. These sperm cells have a row of cilia arranged in a helix at one end and have retained their motility. When they are released, they are in proximity to the archegonium, and one enters to fertilize the egg. The zygote germinates to produce an embryo sporophyte that temporarily stops development after a certain stage and remains in a dormant condition until the environment is favorable for seed germination.

It should be noted that in *Ginkgo* a new means of transporting the sperm to the vicinity of the egg has been attained. The pollen grain (the very young microgametophyte) is carried in its entirety to the ovule, settling in a position near the megasporangium. There is absolutely no dependence upon water as a medium of getting the sperm to the egg. Wind transports the entire gametophyte in the form of

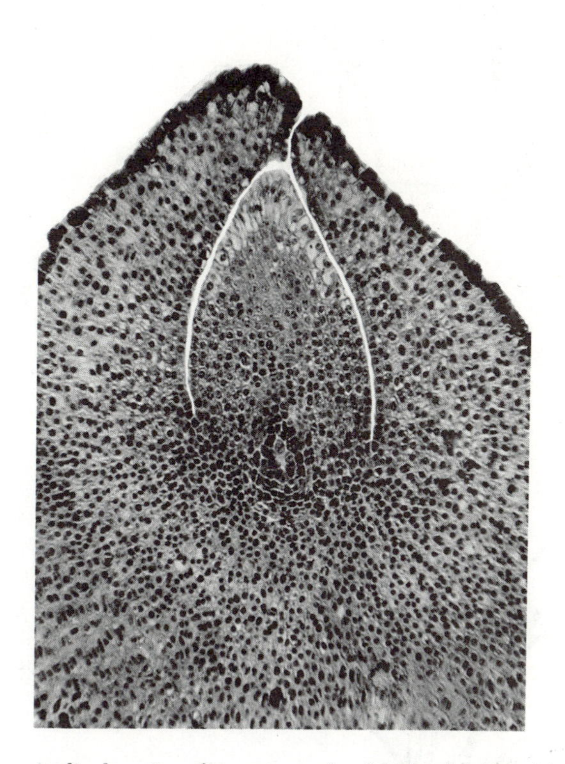

Fig. 4-17 *Longitudinal section of a young ovule of* Ginkgo biloba, *with a megaspore mother cell (large cell near center) within the fleshy nucellus that is completely surrounded by the integument.* ×75. *(Photograph courtesy C. L. Lee.)*

the pollen grain. In *Ginkgo,* even though the sperm cells are carried directly to the archegonium by means of the pollen tube, vestiges of an earlier mechanism of motility have been retained in the form of cilia. Only in *Ginkgo* and members of the division cycadophytes have motile ciliated sperm been retained in the seed plants. In other seed plants all evidence of the former aquatic environment of the sperm cell has been lost; sperm in those plants may be motile, but they move by means of amoeboid movements.

In summarizing the life cycle of *Ginkgo* it may be useful to compare it again with the generalized life cycle presented for all vascular plants. A sporophyte (the *Ginkgo* tree—in this instance, two kinds) produces sporangia (pollen sacs or microsporangia in the pollen-producing tree; megasporangia, *within integuments,* on the ovule-bearing tree). Within the sporangia, spores are produced after meiosis (many in the microsporangia; only four—one functional—in the megasporangium). The spores upon germinating produce gametophytes (the pollen grain represents an early stage in development of the microgametophyte, and further development occurs within the ovule where the pollen grain produces a pollen tube and six cells; the female gametophyte develops within the megaspore). Gametophytes produce gametes (only two sperm cells by the microgametophyte; several eggs by the

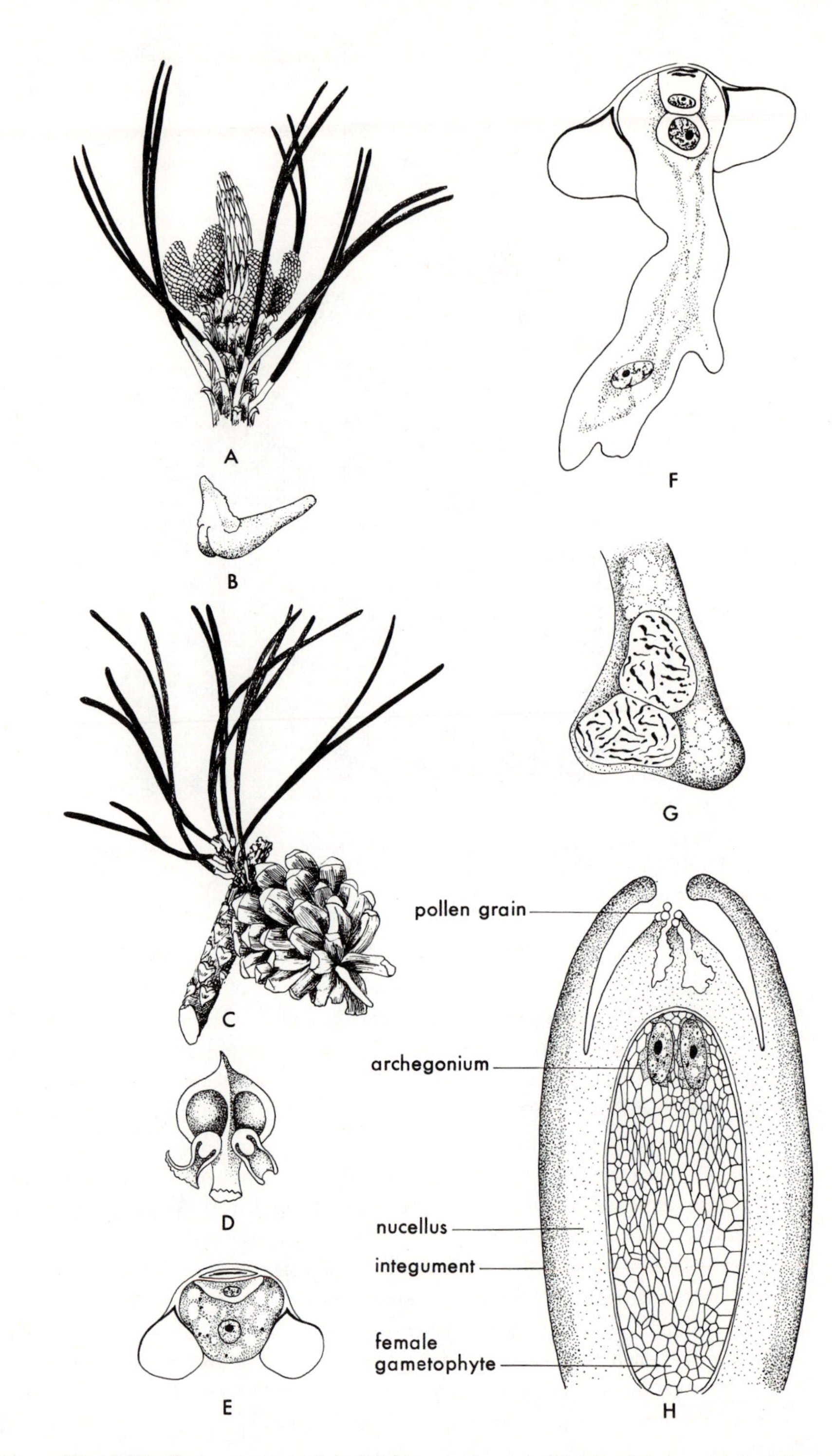

Fig. 4-18 *Various aspects of the life history of a pine. (A) Tip of shoot, with leaf clusters and pollen cones. (B) Microsporophyll, with elongated pollen sacs on the lower side. (C) Shoot, with woody seed cone attached. (D) Young seed cone scale, with two ovules*

megagametophyte), and a sperm cell unites with an egg cell to produce a diploid zygote, the start of the new sporophyte phase.

Reproduction in Conifers

Reproduction in pines and other related conifers is generally similar to that in *Ginkgo* but is somewhat more specialized in a few respects. Unlike *Ginkgo,* most pine trees have both pollen-bearing structures and seed cones on the same plant. Pollen cones consist of an axis on which are arranged small, closely packed microsporophylls (modified leaves bearing microsporangia) in a tight helical arrangement (Fig. 4-18A). Each microsporophyll has on its lower side two elongated sacs in which the microsporocytes undergo meiosis to produce hundreds of microspores (Fig. 4-18B). As in *Ginkgo,* the microspore nucleus divides before pollen is shed, and it is finally released in the binucleate stage. Pine pollen grains and those of certain other conifers are unusual in having two hollow bladders attached to the body of the grain (Fig. 4-18E). These undoubtedly increase their buoyancy in air currents. Pine pollen is produced in an excessive abundance; more is produced than will ever reach ovules. This situation is to be expected among wind-pollinated plants for there is no way to ensure the path the pollen grains will take when transported by air.

Megasporangia in pine are produced in pairs on the upper sides of woody cone scales of the seed cone. The familiar pine cone with helically arranged woody scales on a central axis represents the seed-bearing part of the sporophyte (Fig. 4-18C, D). The megasporangia are enclosed by an integument, with a micropyle left at the end of the ovule directed toward the cone axis (Fig. 4-18D). Within the nucellus (megasporangium) is one megasporocyte; after meiosis of this cell, four haploid megaspores are produced, three of which abort. The surviving megaspore enlarges as the gametophyte develops within it. Archegonia are produced at the end of the gametophyte nearest the micropyle (Fig. 4-18H).

Pollen grains are blown to the ovules and adhere to a sticky pollination drop at the end of the micropyle. As the droplet dries, pollen grains are withdrawn into the micropyle to the vicinity of the nucellus. Here the pollen grain germinates, with a pollen tube penetrating the nucellus and growing toward the megagametophyte (Fig. 4-18F). Of the six cells produced by the male gametophyte, two are sperm cells (Fig. 4-18G). One of these fertilizes the egg in an archegonium to initiate the new sporophyte phase. The resulting embryo enters a resting period after a certain amount of development. The structure that develops after fertilization—consisting of an integument (part of the sporophyte phase) containing the fleshy megagametophyte (haploid gametophyte phase) and within that the new embryo (new sporophyte

---

(sectioned longitudinally) on the upper surface. (E) Pollen grain. (F) Germinating pollen grain, showing young male gametophyte. (G) Tip of a pollen tube, with two enclosed sperm cells. (H) Diagrammatic longitudinal section of an ovule prior to fertilization. [(A) to (C) and (H) redrawn from General Botany for Colleges *by Ray E. Torrey, Copyright, 1922, 1925, 1932 by R. E. Torrey. Reproduced by permission of Prentice-Hall, Inc. (E) to (G) redrawn from J. M. Coulter and C. J. Chamberlain,* Morphology of Gymnosperms, *University of Chicago Press, 1910. (D) adapted from various sources.]*

phase)—is a seed. The fleshy tissue of the megagametophyte in which food is stored is sometimes called *endosperm,* but it is not homologous with the food-storing tissue of flowering plants. This food is utilized by the growing embryo during germination of the seed. Its origin is quite different from that of the endosperm of angiosperms.

In pine the sperm cells show no trace of the ancestral aquatic phase in the life cycle of seed plants. Again, as in *Ginkgo,* water plays no part in transporting the sperm to the egg. Instead the microgametophyte is extremely reduced, and the entire microgametophyte is transported to the megasporangium by some other means, in this particular case, air currents. Furthermore, the sporophyte phase of the life cycle is extremely well-adapted to the terrestrial environment, and while generally only one embryo is produced by one megagametophyte, extremely high numbers of gametophytes can be produced by the massive sporophyte phase—the pine tree.

It is simple to show that the pattern of the life cycle of a conifer is nothing more than a modification of the basic life cycle of all vascular plants. A pine tree represents the diploid sporophyte phase. On it are produced sporangia (microsporangia in pollen cones; megasporangia enclosed within integuments on woody scales of the seed cones). Reduction division of sporocytes within the sporangia results in formation of haploid spores (microspores in pollen sacs; megaspores, only one of which remains functional, in each nucellus). Spores germinate to produce the gametophyte phase. Microspores become pollen grains, and after transport to the ovule the microgametophyte continues development in the form of a tube growing through the nucellus; a megaspore produces a fleshy gametophyte within the megaspore membrane, nucellus, and integument. Gametophytes produce sex cells (two sperm cells in each pollen tube; several archegonia, each with one egg, in the megagametophyte). Fertilization occurs when two gametes unite to form a diploid zygote, the beginning of the new sporophyte phase.

ORIGIN OF THE SEED HABIT IN VASCULAR PLANTS

The series outlined on preceding pages—indicating the increasing amount of reduction in the role of the gametophyte phase and the evolution of the heterosporous habit to an extreme (where only one large functional spore is produced by the megasporangium)—represents steps involved in the origin of the seed. Evolutionary stages leading to the development of the seed habit from a non-seed-bearing plant involve the following: (1) evolution of heterospory, or spores of two kinds (perhaps there may have been no actual initial difference in size and structure between a mature microspore and a mature megaspore, but there certainly was a functional difference); (2) reduction, eventually, to one in the number of spores in a megasporangium; (3) development of the megagametophyte within the megaspore; (4) retention of the megaspore within the sporangium; (5) envelopment of the sporangium by some kind of integumentary system (seed coat).

Heterospory among vascular plants is actually very old; a number of Devonian plants show it. Even in these plants the number of megaspores per sporangium is smaller than the number of microspores in a microsporangium. Further reduction in the number of megaspores is also known in fossil plants, as well as in the present-day *Selaginella,* where four megaspores per sporangium are frequent, and where in some instances there are even fewer than that. Certain Carboniferous and Permian vascular plants, including members of both the lycopods and horsetails, have sporangia with but a single functional megaspore in each.

In *Selaginella* the megagametophyte develops almost entirely within the spore wall, although there is a rupture of the spore membrane. It appears that in the fossil megaspores just mentioned gametophyte development occurred almost completely, if not entirely, within the megaspore membrane.

Retention of the megaspore within the sporangium is a feature that must precede the seed. In fossil megasporangia mentioned (those in the lycopods and horsetails in which there is only one functional megaspore per sporangium) there is good evidence that megagametophytes developed within the megaspores and that the spores were not shed. Isolated megaspore membranes are found in coal and other sediments, but most likely the sporangium had decomposed, while the thick, resistant membrane survived the preservational process. Furthermore, many specimens of intact *Lepidocarpon* sporophylls with spores still within the sporangia are found isolated, suggesting that the whole sporophyll was shed as a unit, not simply the megaspore.

In true seed plants, however, the megaspores are always retained within the nucellus. In addition, the gametophytes are completely confined within the megaspore membrane.

The problem of the evolution of an integument around the sporangium has been a perplexing one. In *Lepidocarpon,* mentioned earlier, an integumentary system of a kind was effected by the enveloping of the megasporangium by the two lateral edges of the sporophyll on which it rests. Although this is an integumentary system that functioned in a protective manner as do typical seed integuments, it is generally agreed that integuments of seeds in the seed plants originated in some other way.

Recent discoveries in rocks of Lower Carboniferous age in Scotland have shed considerable light on the problem. They have revealed certain seedlike bodies in a stage of evolution just preceding that of well-established seeds. One of these seedlike bodies, *Genomosperma kidstoni,* is nothing but a small, slender, elongated nucellus (megasporangium) with pollen-receiving modifications at the apex (Fig. 4-19A). At the base of the sporangium, where it is attached to a slender stalk, are eight elongated fingerlike processes that loosely envelop the nucellus. They really cannot be called integuments as we know them, for the sporangium is not actually covered by them. Furthermore, pollen landed on the nucellus directly, something that does not happen in typical seeds. Yet it would not be unreasonable to consider the fingerlike processes as integument precursors. In fact, a second species of *Genomosperma, G. latens,* shows a partial fusion of these eight processes at the base, indicative of a later stage in the evolution of the integument (Fig. 4-19B). Other seedlike bodies from the same or similar deposits show an almost completely fused and enveloping integument, the original component parts being evident only as lobes at the apex of the seed (Fig. 4-19C–E). In most highly developed integumentary

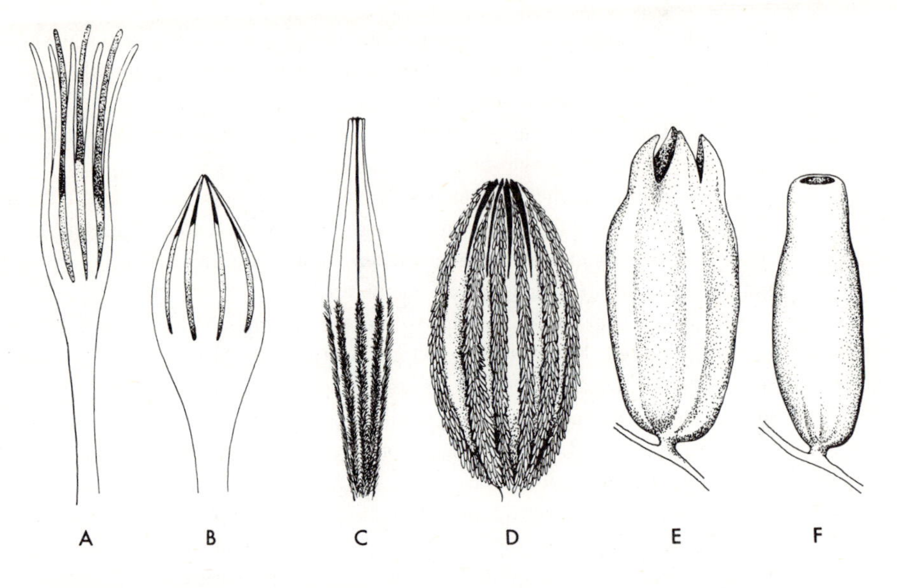

Fig. 4-19 *Fossil seeds or seedlike structures from the Paleozoic Era.* (*A*) Genomosperma kidstoni. (*B*) G. latens. (*C*) Salpingostoma dasu. (*D*) Physostoma elegans. (*E*) Eurystoma angulare. (*F*) Stamnostoma huttonense. (*Redrawn from H. N. Andrews,* Science, *vol. 142, p. 927, 1963. By permission.*)

systems, there is no vestige of the fingerlike tips (Fig. 4-19F). In many seeds, therefore, the integument may be interpreted as being the product of the fusion of a number of slender branches arising at the base of the nucellus and extending over it.

The function and adaptive value of an integument, or seed coat, appear to be obvious. In seeds of most seed plants the zygote germinates to produce an embryo that undergoes a dormant stage. Germination occurs only after certain conditions within the seed and in the environment have been fulfilled. These conditions vary among different kinds of seeds. The integument serves as a protective device, functioning largely for mechanical protection of the embryo. In some seeds there also seems to be an efficient "gasproofing" function of the seed coat that prevents oxygen from entering. In fact, the protective function of some seed coats has been carried to such an extreme that these seeds normally will germinate only after the seed coat has been somehow weakened, either mechanically or chemically.

In some flowering plants, as will be seen in the following section, there has been a reinforcing of the protective covering of a seed by the fusion of a fruit wall to the seed coat.

ANGIOSPERM REPRODUCTION

In the flowering plants there has been an extreme reduction of the gametophyte phase and a corresponding elaboration of the sporophyte phase in the reproductive process. In the angiosperms, seeds are not borne naked but are enclosed within a specialized structure called a *carpel* or *pistil,* which at maturity becomes the fruit wall. The number of seeds borne within a carpel varies considerably among the species of flowering plants, ranging from one to hundreds. (Although in some naked-seeded plants such as certain conifers it appears that the seeds are not borne exposed to the air but are really tightly enclosed by component parts of the cone, the fact that pollen lands directly on the ovules makes them typically gymnospermous. This obviously suggests that at some stage in cone development, the ovules were not completely enclosed and were readily exposed to the pollen grains.) Pollen in the flowering plants does not land directly on the ovule but on a special pollen-receptive organ of the enclosing carpel, usually at the tip, called the *stigma.*

In spite of specializations, the angiosperm life history is basically identical with that of any other vascular plant. The sporophyte phase that is the conspicuous and familiar part of the life cycle (for example, a magnolia tree, a sunflower plant, a maize plant, a lilac bush, an aspen tree) bears flowers in which are produced the sporangia. Microspore-producing organs are the structures called *stamens,* a typical one of which consists of a slender, threadlike *filament* and a swollen, pollen-producing *anther* (Fig. 4-20). Within the anther are microsporocytes that undergo meiosis to produce uninucleate, haploid microspores (Fig. 4-21). Soon after microspore production the multicellular phase of the microgametophyte stage is initiated by the mitotic division of the microspore nucleus. Two cells are formed—a tube cell and a generative cell. At this stage the small sporelike body, called a pollen grain, is shed.

In the same flower, typically—though not necessarily (or necessarily even on the same plant)—are produced carpels, within which are ovules (Fig. 4-20). Each ovule, of course, consists of a nucellus (megasporangium) with an integumentary system. One sporocyte within each nucellus undergoes meiosis to produce four haploid megaspores (Fig. 4-21). What happens after this stage varies considerably among the angiosperms. Typically, only one megaspore is functional, the other three aborting. In some species more than one may remain functional. Whatever the situation, however, only one megagametophyte is produced.

In this "typical" situation in which only one megaspore functions, further events that follow include the division of the megaspore nucleus into eight; around six of them walls are formed, with the other two occupying the central region of what is actually a large, seventh cell (Fig. 4-21). This multicellular female gametophyte is called an *embryo sac* in the angiosperms. Generally the six small cells arrange themselves so that three are at each end of the embryo sac, with the two additional nuclei near the middle. One of the three cells toward the micropylar end is the egg cell, and it is flanked by the other two cells called *synergids.* Before fertilization the two nuclei in the center of the embryo sac fuse to produce a diploid fusion nucleus (Fig. 4-22).

At the time of the opening of the anther (microsporangium), pollen is released and somehow transferred to the pollen-receptive surface of the carpel in the same or different flower. Wind is the usual vector among gymnospermous plants and, to a

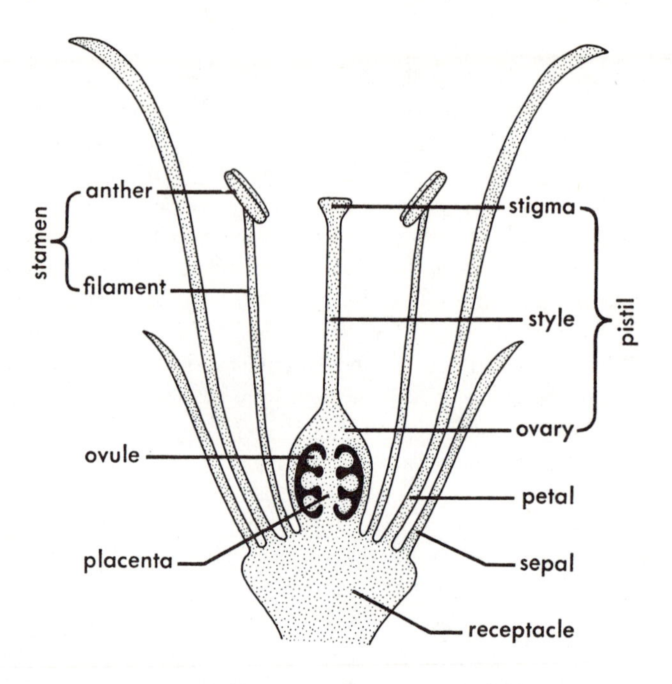

Fig. 4-20 *Diagrammatic longitudinal section of an angiosperm flower.* (*From* Botany, Fifth Edition, *by C. L. Wilson, W. E. Loomis, and T. A. Steeves, Copyright 1952, © 1957, 1962, 1967, 1971 by Holt, Rinehart and Winston. Reprinted by permission of Holt, Rinehart and Winston.*)

large extent, among angiosperms. More important, however, is the role of pollinating insects, which have effected some extremely refined pollination systems.

Because the pollen grain is transferred to the carpel of the flower rather than directly to the ovules, as among the gymnosperms, there is usually a considerable distance between the egg of the female gametophyte and the spot where the grain lands on the carpel. The pollen grain, however, germinates and grows through the carpellary tissue in the form of an elongated tube (Fig. 4-21). Tube growth is directed toward the ovules and eventually a pollen tube penetrates the micropyle of the ovule. While the tube is growing, the generative cell divides into two sperm cells; the total number of cells in a microgametophyte is three (Fig. 4-21).

Sperm cells are released from the tip of the tube into one of the synergids; one of the sperm cells then migrates to the adjacent egg and unites with it to form the diploid zygote, the beginning of the new sporophyte phase. The second sperm is also functional; its nucleus unites with the fusion nucleus in the middle of the embryo sac. In a "typical" embryo sac such as that just described, this nucleus after fertilization would be *triploid* (that is, it would have three times the haploid number of chromosomes). After many divisions of this triploid nucleus a multicellular food-storing tissue is formed. This tissue is called endosperm. At maturity, therefore, a seed consists of an outer seed coat, an embryo (new sporophyte phase), and an endosperm (stored food). The nucellus, as in most mature seeds, has been crushed

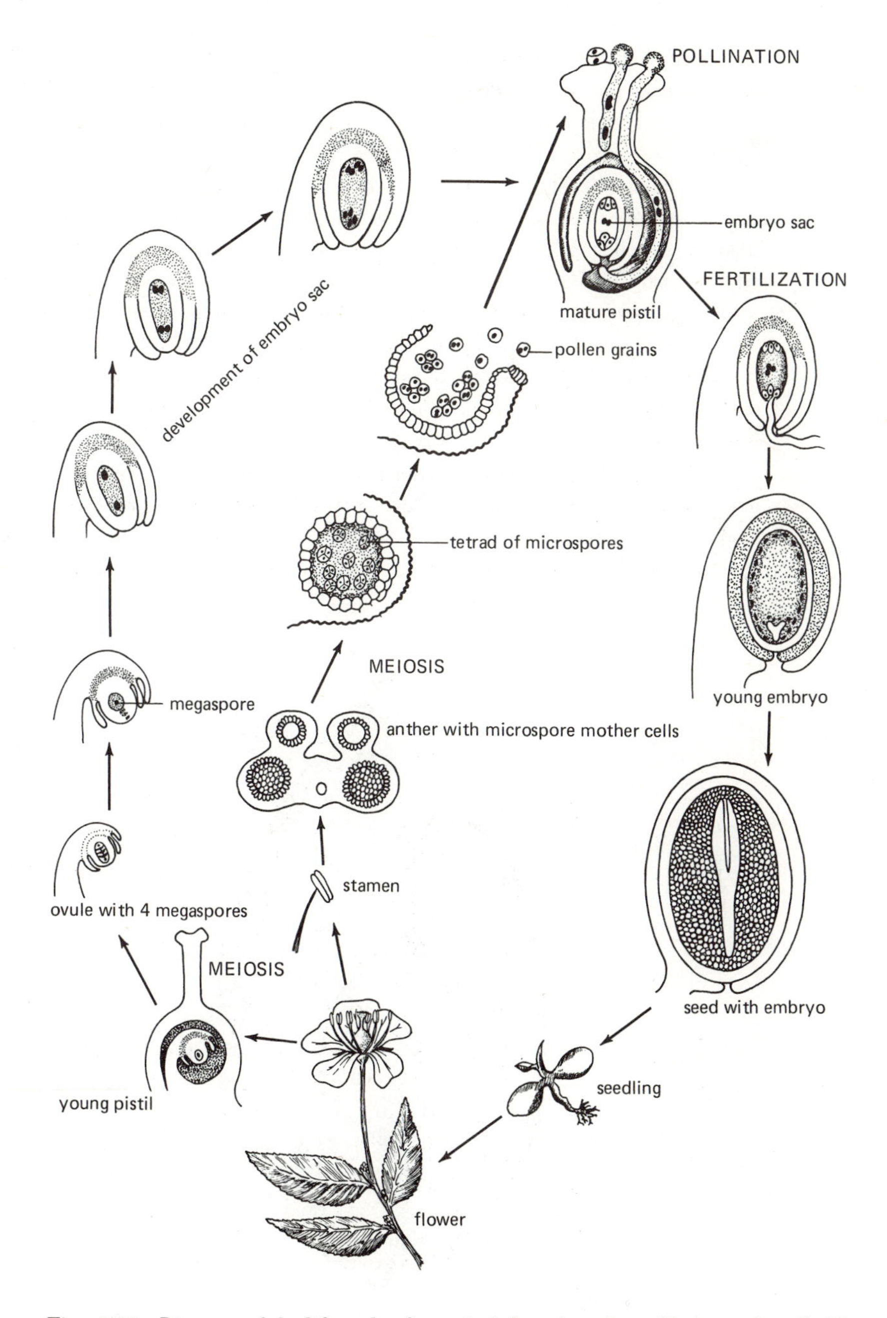

Fig. 4-21 *Diagram of the life cycle of a typical flowering plant. (Redrawn from E. W. Sinnott and K. W. Wilson,* Botany, Principles and Problems, *McGraw-Hill Book Company, 1963. By permission of the publishers.)*

and is no longer discernible. Endosperm provides nutrients for the developing embryo as the seed germinates; in some plants, however, food from the endosperm is digested and absorbed by the embryo prior to germination, in which case no endosperm is recognizable in the mature seed.

Although having the same function as the food-storage tissue in naked-seeded plants, angiosperm endosperm has a quite different origin. Gymnosperm food-storing tissue is simply the swollen female gametophyte in which food is stored; in the angiosperms it is derived from divisions of a cell that was formed by fusion of two female gametophyte cells and a sperm cell. The phenomenon among flowering plants whereby one sperm unites with an egg and a second unites with two other cells in the female gametophyte is called *double fertilization.* Double fertilization occurs only among angiosperms and is one constant criterion that distinguishes the flowering plants from all other vascular plants.

Through the preceding report of angiosperm reproductive events (summarized in Fig. 4-21), advanced though they are, it is still possible to see how they are simply modifications of the basic life cycle for all vascular plants. On the sporophyte phase (the tree, shrub, herb, and so forth) are produced sporangia (microsporangia in the

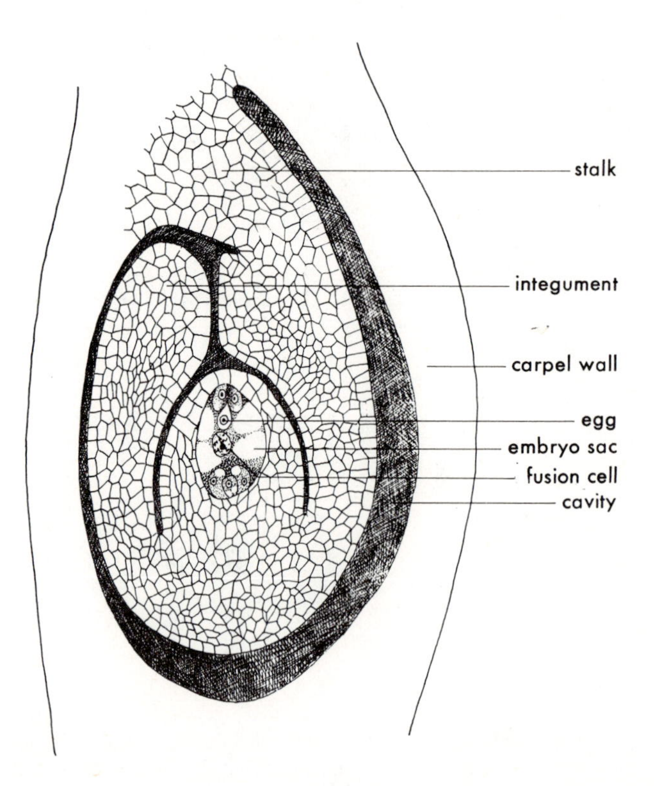

Fig. 4-22 *Longitudinal section of part of a carpel of* Anemone patens, *with one enclosed ovule. (Redrawn from A. W. Haupt,* Plant Morphology, *McGraw-Hill Book Company, 1953. By permission of the publishers.)*

form of anthers; megasporangia surrounded by integuments and contained within a carpel). Meiosis of sporocytes in the sporangia results in formation of haploid spores (microspores in the anthers; megaspores, with generally only one remaining functional, in the nucellus). Spores germinate to produce the haploid gametophyte phase (first the two-nucleate pollen grain, then the pollen tube with three cells at maturity, and finally eight cells of the female gametophyte—the embryo sac). Gametes produced by the gametophyte phase unite (a sperm cell unites with the egg) to produce a diploid zygote, the beginning of the new sporophyte phase.

SUMMARY Throughout this discussion of reproduction in vascular plants diversification of life cycles has been presented, but an effort has been made to indicate how all of this diversity is simply a series of elaborations on the same theme. While there are obvious differences in the particulars, it becomes a relatively simple matter to remember and understand these life cycles if the basic underlying theme is kept in mind.

In the life cycles just discussed, a number of evolutionary changes have occurred. The first is the increased specialization of the sporophyte. In the most primitive vascular plants, sporangia are borne exposed to the air, and only one kind of spore is produced. Progressively more advanced heterospory developed, whereby the numbers of megaspores were reduced. At the same time, food-storage capacity within the megaspore was increased. The end result was a final reduction to one functional megaspore, with three of the spores produced as a result of meiotic division of the single megasporocyte aborting. Eventually protective integumentary systems were developed around the megasporangia—part of the sporophyll in *Lepidocarpon,* but sterile fingerlike branches among some of the true seed plants. These sterile branches became progressively more fused to form an integument. Finally, in the angiosperms these ovules (integumented megasporangia) became enclosed within another structure, the carpel.

Along with the progressive specialization of the sporophyte phase there occurred a corresponding progressive decrease in the size and degree of development of the gametophyte phase. Primitive land plants had free-living gametophytes that were often photosynthetic, attached themselves to the soil by means of absorbing hairs, and produced both antheridia and archegonia. As heterospory in the sporophyte phase evolved, two kinds of gametophytes were formed as a result of the germination of two kinds of spores. In these instances the gametophyte phase was largely retained within the spore, and only part of it was exposed in certain megagametophytes. The male gametophyte was reduced essentially to a single antheridium, composed of a wall surrounding a number of sperm cells. Further evolution resulted in a still more pronounced reduction and complete enclosure of the megagametophyte within the spore wall. The number of cells within the male gametophyte was also progressively reduced. In fact, not even an antheridium can be recognized in the several-celled gymnosperm microgametophyte, while in the angiosperms three cells are the total in the entire microgametophyte phase. Angiosperm megagametophytes are also extremely reduced, with generally only seven cells involved.

The gametophyte generation among the most highly evolved vascular plants has been streamlined down to almost the barest essentials—the gametes. Furthermore, there is absolutely no dependence upon water for the transfer of sperm to the egg. Practically all of the development of the male gametophyte takes place in the vicinity of the ovule (with its megasporangium and contained spore and megagametophyte). The entire male gametophyte phase is transported by wind, insects, other animals, and sometimes by water (however, not sperm cells alone, but the entire microgametophyte) close to the ovule, and sperm cells are brought even nearer by the development of a tube that grows practically to the egg.

All of these modifications serve as adaptive features in a land environment. The gametophyte of primitive vascular plants is geared to an aquatic existence; progressive sporophyte modifications serve for better adaptation to land. Accompanying reduction of the gametophyte occurs until it is the more insignificant of the two phases in the most highly specialized land plants.

It is obvious that in spite of the better adaptive features of higher vascular plants—seed plants in particular—the more primitive vascular plants and even the nonvascular land plants (bryophytes) live quite successfully in environments generally dominated by the seed plants. The reason is that they occupy niches to which their less specialized means of reproduction are quite well adapted. Furthermore, these plants often grow in moist, shaded places, probably because they are more tolerant of a decreased light intensity than are most seed plants. Ferns and lycopods are quite successful in their respective habitats and will probably continue to exist, in spite of the preponderance of seed plants, for some time to come. On the other hand, in their present state they could not possibly be as successful in as wide a range of environments as the seed plants are.

FURTHER READING

Andrews, H. N., "Early Seed Plants," *Science,* vol. 142 (1963), pp. 925–937.

Banks, H. P., *Evolution and Plants of the Past.* Belmont, Calif.: Wadsworth Publishing Company, 1970.

Bold, H. C., *Morphology of Plants.* 3d ed. New York: Harper & Row, 1973.

Delevoryas, T., *Morphology and Evolution of Fossil Plants.* New York: Holt, Rinehart and Winston, 1962.

Editors of Time-Life Books, *Life before Man.* New York: Time-Life Books, 1972.

Foster, A. S., and E. M. Gifford, Jr., *Comparative Morphology of Vascular Plants,* 2d ed. San Francisco: Freeman, 1974.

Ray, P. M., *The Living Plant,* 2d ed. New York: Holt, Rinehart and Winston, 1972.

Sporne, K. R., *The Morphology of Gymnosperms.* London: Hutchinson University Library, 1969.

Sporne, K. R., *The Morphology of Pteridophytes.* London: Hutchinson University Library, 1970.

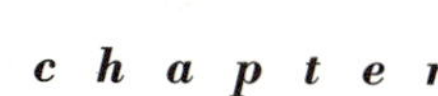

chapter 5

Flowering Plant Evolution

The flowering plants, or angiosperms, are currently the most widespread of land vascular plants, even though they are the most recently evolved. Furthermore, they are the most important group of plants in our economy. It is therefore natural for people to be especially concerned about the origin and subsequent evolution of the division.

GEOLOGIC EVIDENCE OF ANGIOSPERMS

Origin of the group in time is still somewhat vague. There are a number of reports of angiospermlike plants before the Cretaceous Period (Mesozoic Era), but the first certain ones appeared then. This does not mean that reports of pre-Cretaceous angiosperms should be ignored; it simply means that more evidence is necessary to prove that the plants in question were actually flowering plants.

One report of such pre-Cretaceous plants is interesting because of the age of the specimens involved. From Triassic rocks of Colorado and Texas have been collected impressions of leaves resembling those of palm trees (Fig. 5-1). Stem fragments are also known, suggesting a short, squat habit of the plants that bore these leaves. There would be no hesitation in calling these palms if they had been found in Cretaceous or Tertiary rocks. Their much greater age, however, tends to make paleobotanists cautious about interpreting them unequivocally as angiosperms. Even though it has been established that the leaves are, indeed, like those of palms, nothing is known of the reproductive system of these plants, and such a plant may simply have been one that had attained a more advanced evolutionary stage in its vegetative features than in its method of reproduction. Such transitional types of plants should be expected in the fossil record, for it is almost universally recognized that various organs of plants and animals may evolve at independent rates, and that the first "angiosperm" probably did not have all of the characteristics typical of

Fig. 5-1 *Impression of a palmlike plant,* Sanmiguelia, *from the Triassic of Colorado.* $\times 0.3$.

angiosperms. For example, certain angiosperm families mentioned earlier have a primitive type of vesselless wood resembling the wood of gymnosperms.

Until quite recently, remains of plants from the early part of the Cretaceous Period have been identified as leaves of certain angiosperm families considered to be primitive (for example, the magnolia and laurel families). Critical examination of the fossil leaves, especially of the fine venation pattern, show that the plants cannot be assigned to any existing family of angiosperms and that most likely they represent truly primitive and probably extinct forms (Fig. 5-2).

Progressively higher in the geologic column it becomes evident that the venation pattern of angiosperms changes from a simple, somewhat irregular type of vein system to a more highly organized pattern of branching and anastomosing.

WHAT IS AN ANGIOSPERM?

It may be appropriate to ask the question, "What is an angiosperm?" at this time. Angiosperms are commonly called the "flowering plants," and we might expect that the occurrence of flowers in the group would be sufficient to distinguish them from all other plants. Some angiosperms, however, do not have flowers as we might generally think of them. There may simply be a solitary carpel, or perhaps only a cluster of stamens, with no resemblance whatever to the usual concept of a flower. Duckweeds, for example, are minute aquatic plants in which the flower is hardly recognizable. Thus the use of the criterion of whether or not a vascular plant has flowers cannot work as a means of differentiating angiosperms from other plants until it has been decided just what a flower is.

It is frequently said that the presence of a closed carpel surrounding ovules is characteristic only of the angiosperms. Although this seems to be a feature found in practically all of the flowering plants, in some cases it is difficult to know just how "closed" a carpel should be to qualify as one. Certain angiosperms have incompletely closed carpels, with ovules actually exposed to the outside air. Pollen grains, in fact, may actually be transported to a position *within* the carpel and adjacent to the ovules, a phenomenon normally associated with gymnospermous plants.

It is usually stated that angiosperms as a group have vessels in the wood, and that other groups generally lack them. However, as more anatomical work is done on some of the so-called "lower" vascular plants, more evidence of vessels is coming to light. Certain gymnosperms (for example, members of the Gnetophyta) are also known to possess vessels. And, as indicated earlier, some of the angiosperms themselves actually lack vessels in the wood; some apparently had wood that never evolved to that state, and others appear to have had vessels at an earlier time and to have lost them subsequently.

For some time the seven-celled embryo sac was considered to be a good means of distinguishing angiosperms from other seed plants. The gametophyte is certainly the most reduced of all seed plant gametophytes, and seven cells in the megagametophyte appeared to have been the ultimate in reduction. More detailed work on angiosperm reproduction, however, indicates that there are actually a great variety of kinds of embryo sacs among the flowering plants, and it is not possible to select only

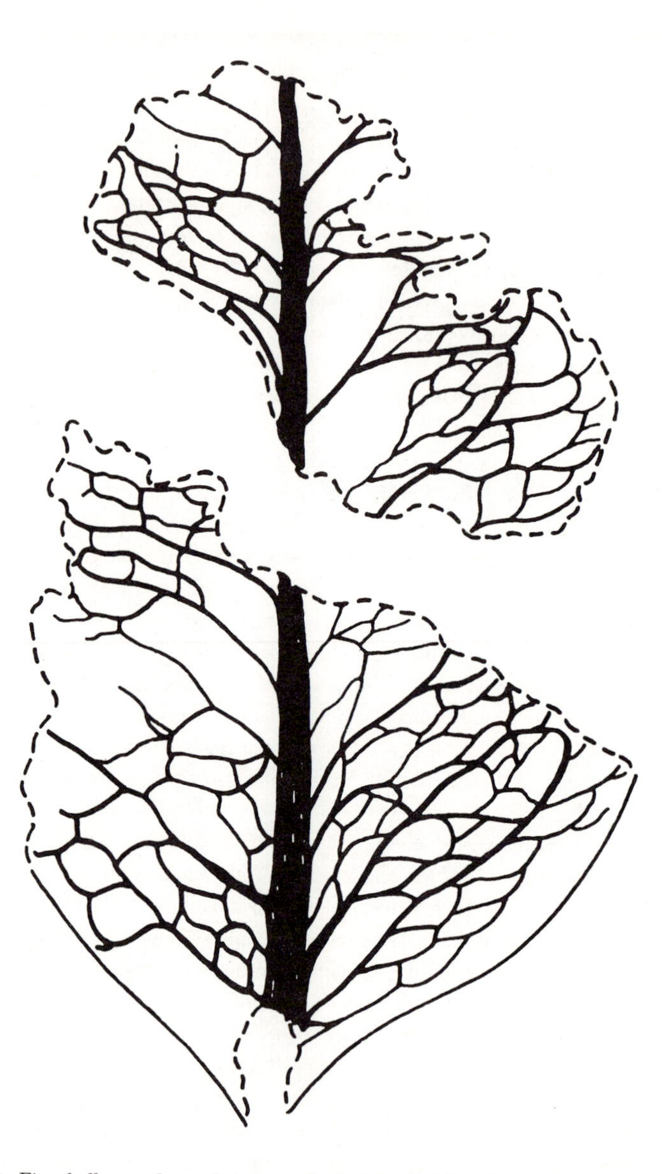

Fig. 5-2 Ficophyllum, *a lower Cretaceous leaf with primitive, disorganized venation.* × *1. (From J. A. Doyle and L. J. Hickey, in* Origin and Evolution of Angiosperms, *C. B. Beck (ed.), Columbia University Press, 1976.)*

one type as representative of all angiosperm megagametophytes. Many embryo sacs have much more than seven cells; some have fewer. Certain embryo sacs have origins from more than one megaspore; others owe their origin to a single functional spore. Thus embryo sac structure cannot be used as a means of tying together all of the angiosperms into one group.

One feature, however, appears to have some significance. This is the phenomenon of double fertilization, whereby two sperm cells are functional—one fertilizing the egg, the other uniting with the fusion cell of the female gametophyte to produce the cell that initiates endosperm formation. No other vascular plants have a comparable feature.

It should be obvious that with the difficulty in agreeing upon criteria that make angiosperms different from all other vascular plants, it would be even more difficult to recognize among the fossil plants the first appearance of angiosperms. Because a certain fossil has leaves resembling those of a palm is not proof that it is an angiosperm. Similarly, finding a plant with wood that lacks vessels in rocks of Mesozoic age is not a true indication that the plant is not an angiosperm.

It is quite likely that the earliest fossil angiosperm will be different from contemporary angiosperms. It may have a closed carpel, for example, but there may be nothing resembling a flower. Or it may have flowers, with leaves and wood quite different from those of modern angiosperms. With the understanding that not all organs evolve at the same rate, it would be extremely difficult to decide at which point in evolution the gymnospermous condition (which we assume was ancestral to the angiospermous condition) changed to an angiospermous one. The one criterion that seems to be almost universal among the angiosperms, double fertilization, unfortunately is one that may never be recognized from fossil material.

Because the angiosperms are so significant a group of plants, many workers have concerned themselves with their possible origin and subsequent evolution. The problem is such a baffling one that practically every botanist has thought about it at least at one time or other.

Evidence from the fossil record seems to suggest a rather rapid period of expansion of the group after its first appearance. Jurassic floras show no convincing angiosperms, but in the Cretaceous there appears to have been a sudden burst of angiosperm evolution and distribution. (Within the Cretaceous Period there was a noticeable increase in abundance and distribution of flowering plants from the beginning of the period to the end.) Of course, this sudden burst does not necessarily preclude a much earlier origin of the group. It is very possible that angiosperms could have originated very early in the Mesozoic Era (some paleobotanists have even suggested late Paleozoic) and might have been too insignificant in numbers and distribution to have been noticed in the fossil record. The sudden expansion of the group later in the Mesozoic was probably due to environmental conditions and inherent adaptive features in these plants coinciding fortuitously. At the present time the angiosperms are the dominant group of vascular plants on the earth's surface (and in some cases living quite successfully under the surface of the water).

FLORAL EVOLUTION

Ideas about the simplest angiosperms and their subsequent evolutionary pathways are based primarily on studies of modern flowering plants. Until more fossils are available, this is the best approach.

A diagram of what may be considered a typical flower and its parts is presented in Fig. 4-20. Not all flowers, however, agree with this diagram in their structural plans. It is generally believed by most workers in this area that the most primitive angiosperm had large flowers, with an elongated floral axis bearing the flower parts in a helical arrangement. Most botanists agree that a flower is morphologically equivalent to a shortened branch system bearing appendages, and it is fairly safe to assume that a flower having most in common with a branch would be very primitive. The number of floral parts is indefinite but fairly large. Parts of the flower are all separated from each other (that is, there is no fusion between parts of the same kind or between one type of flower part and another). It is also generally agreed that flower parts are homologous with leaves, and flowers with parts looking most like leaves are generally considered to be primitive. In other words, leaflike sepals, petals, stamens, and even carpels are found among angiosperm families that are regarded as among the most unspecialized.

Insect versus Wind Pollination

Pollination by insects is also thought to be a feature of the most primitive flowers (though it also appears in many advanced kinds). This statement may appear to contradict earlier remarks that wind pollination in seed plants came before insect pollination among the flowering plants. In the angiosperms, however, wind pollination appears to have been derived, and flowers that are structurally adapted for wind pollination have been altered from a previous type of floral structure suitable for insect pollination. Among the wind-pollinated plants there is generally a type of wood structure that is not found in the most primitive kinds of angiosperm wood. This in itself is not proof that wind pollination followed insect pollination, because it has been suggested already that all organs or parts of a plant need not evolve at the same rate. It could be conceivable that wind-pollinated flowers are more conservative than other parts of the plant and are representative of a primitive floral condition, while anatomical details of vegetative parts of the plant have become more specialized.

There is evidence, however, that certain of the simple wind-pollinated flowers represent a derived and reduced condition from earlier, more complex insect-pollinated flowers. Flowers pollinated by the wind frequently lack either stamens or carpels (stamens and carpels are borne on two kinds of flowers) and possibly some of the outer floral envelope (sepals and petals). Obviously, brightly colored, insect-attracting petals serve no apparent adaptive function when wind is the primary pollinating agent. The former presence of such missing parts in wind-pollinated flowers, however, is often attested to by vascular bundles in the floral axis that appear to be leading to some floral part that is no longer there. In other words, the stamens or petals or other parts in question have been lost, but the vascular system leading to them is still present in a vestigial state. This is probably the most convincing evidence that wind-pollinated flowers have been altered from insect-pollinated flowers. Flowers with all floral parts present (that is, sepals, petals, stamens, and carpels) are considered to represent a condition that preceded one in which some of the flower parts are absent.

Other Floral Modifications

Floral evolution proceeded from a primitive flower structure along a number of different evolutionary pathways, but there were a number of evolutionary trends that seem to have been common to all. For example, there was a marked further telescoping of the floral axis or receptacle, so in the more advanced flowers it was more nearly flattened and less stemlike in appearance. There was also a gradual shift of arrangement of parts from a helical one to a more nearly cyclical one; flower parts were borne in whorls with, for example, the sepals arising at one level, petals at another level, and so forth.

Along with this trend toward cyclic arrangement came the gradual reduction in number of each kind of flower part and a more constant number of parts in the flowers of a given species. For example, a tulip flower almost always has six colored perianth members; six stamens is the usual number, and three carpels (fused together) is the general condition.

There are a number of additional trends, although they are not common to all lines of floral evolution. One trend is the evolution from primitive flowers with a radial symmetry to those having a bilateral symmetry. Usually the evolution of bilaterally constructed flowers is associated with more specialized insect-pollinating mechanisms. In radial flowers an insect may approach the center of the flower in any number of positions; in bilaterally symmetrical flowers there is generally only one position in which the pollinating insect in search of nectar or pollen may reach the center of the flower (Fig. 5-3). Inner flower parts in bilaterally symmetrical flowers are so arranged that the insect touches either the anthers or stigma of the carpel, or both. In that way, pollination is effected virtually every time an insect moves from one flower to another of the same species.

Flowers that have one or more of their parts missing (incomplete flowers) are generally considered to have been derived from flowers with all of their parts (complete flowers). The former presence of parts may be indicated by vestigial vascular bundles. Often flowers may have only stamens or only carpels, and the two kinds of flowers may be borne close to each other on the same plant, be widely separated from each other on a given plant, or even be on two separate plants. Some plants (for example, poplar) may have only pollen-bearing flowers on one tree and only carpel-bearing flowers on another.

Certain flowers have some of their parts fused (for example, petals of a bluebell). These are thought to have been derived from flowers with all of their parts separate. Not only may sepals fuse with sepals, petals with petals, and so forth, but there may be fusion between kinds of floral parts (sepals with petals, stamens with petals).

Position of the carpels on the receptacle and in relation to other floral parts is also quite variable. In a primitive flower the carpels are inserted on the surface of the receptacle and occupy a position slightly above the rest of the flower parts. In certain more specialized flowers the receptacle near the carpel(s) actually grows beyond the position of carpel insertion, and some or all of the rest of the flower parts may arise from a position above that of the carpels. In extreme types the growth of

Fig. 5-3 *Bees visiting bilaterally symmetrical flowers of snapdragon. (Photo by C. F. Hottes. From H. J. Fuller and O. Tippo,* College Botany, *Holt, Rinehart and Winston, 1954.)*

the receptacle around the carpels may produce a cuplike structure, with other flower parts borne along the rim. This cup in some families may fuse against the carpel wall, so that not only are ovules surrounded by a carpel, but there is receptacle tissue as well. A similar end result may be accomplished when the basal parts of the flower around the carpels become fused into a tube that elongates and surrounds the carpel, and often fuses with the carpel wall. This is the situation in an apple fruit, in which the actual mature carpels are the cartilaginous structures within the core, and the fleshy part represents fused bases of sepals, petals, and stamens.

Perhaps a word concerning the actual mechanics of the "fusion" so often referred to is necessary here. In practically all cases cited there is not really a fusion in the sense that parts come together and adhere organically. What really occurs in the ontogeny of fused petals, for example, is the growth initially of separate petal primordia. Subsequent development involves the growth of a continuous ring of tissue that elevates the free petal tips at the end of a continuous tube. Fusion in this sense implies a phylogenetic series of events that are expressed in ontogeny by the development of a ring of tissue. Admittedly, there are examples of ontogenetic fusion, but they are not as common as the type of phylogenetic fusion described here.

Floral Aggregation An extreme advancement in floral arrangement on plants is found among members of the composite family (to which belong such plants as daisies, asters, and dandelions). In the composite group the flowers are compactly situated on a generally flattened end of a stem axis; flowers are frequently small, and the whole assemblage, called a *head,* often takes on the appearance of a single flower (Fig. 5-4). So close is the resemblance that in certain of the composites, the head actually seems to have sepals, petals, stamens, and carpels. Upon closer examination, however, it can be seen readily that what appear to be sepals are in reality reduced leaves compactly arranged around the base of the head. The petallike members are individual flowers, the petals of which have been flattened and fused into strap-shaped structures. The stamens of these peripheral flowers are generally absent, and sometimes even the carpels are missing. These showy, peripheral flowers, called *ray florets,* are obviously insect-attracting devices that serve the same function as do the petals of an individual flower. The innermost flowers, called *disk florets,* generally are quite small, with a

Fig. 5-4 *Sunflower* (Helianthus), *a member of the composite family. (A) Flower cluster at the tip of a branch. (B) Ray floret, with strap-shaped, fused petals. (C) Disk floret, with anther tube projecting beyond tube formed by fused petals. Cluster of pollen grains being pushed out of the anther by the style is visible at the tip. (D) Disk floret in the pollen-receptive stage, with the style split at the tip. (Redrawn from* General Botany for Colleges *by Ray E. Torrey. Copyright 1922, 1925, 1932, by R. E. Torrey. Reproduced by permission of Prentice-Hall, Inc.)*

very much reduced set of sepals and petals. They are usually perfect (both stamens and carpels are present), and function principally in the production of pollen and seeds. These central flowers are the ones to which the pollinating insects are attracted by the peripheral circle of colorful ray florets. The actual pollination mechanism in these composite disk florets is extremely interesting and considerably specialized, and will be discussed with other pollinating devices later in this chapter.

Although aggregation of flowers to simulate a single flower is carried to an extreme in the composite family, this same tendency is present in a number of other families of flowering plants. In the common wild carrot (or Queen Anne's lace) the flowers are clustered in a flat plane, with the peripheral flowers somewhat larger than the rest and with an innermost, central, purplish or maroon-colored flower. Flowering dogwoods also have aggregated flowers that give the appearance of single flowers (Fig. 5-5). The four large white or pink petallike members are not petals at all, but are modified leaves that surround the actual flower cluster. The flowers themselves are rather small and are borne in a tight little head. Again, the bright peripheral leaves are probably correctly interpreted as insect-attracting structures. A parallel situation occurs in the poinsettia plant, commonly blossoming in the winter (Fig. 5-6). The showy red "petals" are really modified and highly colored leaves that surround the flower clusters. In the center are the individual flowers, but the structure of these flowers too is highly specialized. They are not like the flowers in a dogwood cluster, but each of the central flowerlike structures represents an aggregation of greatly reduced individual flowers.

The fact that there are many other instances where aggregation of flowers has occurred (even the common clover plant has heads composed of numerous individual flowers) suggests a very likely adaptive value of this phenomenon. One value that is fairly obvious might be the assurance that more flowers will be pollinated since so many sources of pollen are nearby. The presence of colored or enlarged members at the periphery of the floral cluster (either modified leaves or other flowers) makes insect attraction more likely. Even though the full adaptive significance of aggregation may not be known, the fact that it is so common an occurrence among the flowering plants makes it seem a highly successful feature.

The principle of aggregation seems to have been carried to an even more advanced degree in some composites; for example, in the globe thistle, the individual compound heads of flowers are themselves aggregated into a larger, tightly packed cluster.

EVOLUTION OF POLLINATING MECHANISMS

Evolution of flower structure is closely tied to the evolution of pollinating mechanisms. Among flowers pollinated by animals, those with insect pollen vectors have frequently attained extremely specialized, often bizarre, structural adaptations.

Some of the angiosperms that are considered to belong to primitive families

Fig. 5-5 *Flowering dogwood* (Cornus florida). *Showy, petallike structures are actually modified leaves surrounding the clusters of small flowers.* (*With permission of the National Park Service.*)

are pollinated by beetles, and in these flowers there is actually little structural specialization that would help ensure pollination with any high degree of frequency. Pollination is effected simply by beetles or other chewing insects that feed on floral parts; as they move from one flower to the next, pollen is inadvertently transported from one flower to the carpels of others. These flowers generally have an abundance of floral parts, so the chewing away of many stamens or carpels by an insect may still leave many others unaffected and functional.

This type of pollinating mechanism is not the most common among angiosperms, however. Flowers pollinated by other orders of insects (such as Hymenoptera and Diptera) often have structural features that make pollination almost a certainty with each insect visit. Typically, in such plants, accessory floral organs that secrete nectar (*nectaries*) attract insects that brush against the stamens and carpels as they feed. Some pollen grains remain adhering to the body of the insect, and as it feeds on another flower, this pollen is deposited against some of the carpels, while additional pollen is picked up.

Some insects (for example, the common honeybee) actually collect pollen; as

Fig. 5-6 *Poinsettia* (Euphorbia pulcherrima), *with showy, colored leaves surrounding reduced flower clusters. (Redrawn from William H. Brown,* A Textbook of General Botany, *1925, published by Ginn and Company.)*

each insect picks up pollen from the stamens, some of the pollen it has collected previously is brushed against the stigmas of carpels.

In many radially symmetrical flowers, however, there is no guarantee that a visiting insect will pick up pollen or deposit it on a carpel with each visit. An obvious advance, and what appears to be a more foolproof pollinating system, as explained earlier, is found among flowers that have bilateral symmetry.

Members of the composite family are interesting because of the aggregation of flowers that make it possible for insects to visit many individual flowers in a short time, as well as for an unusual pollination mechanism. Disk florets of composites typically have five stamens; the filaments are free, but the elongated anthers are fused into a tubelike structure (Fig. 5-4C). When the anthers split at maturity to release the pollen, it is shed into the inner part of the tube. During the maturation process of the flower the style is elevated through the tube and is covered at the surface by hairs. Thus, as the style elongates and the filaments of the stamens contract, the style functions much in the manner of a bottle brush, pushing out the pollen grains and exposing them to the air beyond the distal opening of the anther tube. Insects then carry off the pollen and deposit it on stigmas of other flowers. At the time the stigma is pushing out the pollen from the inside of the tube, it has not matured as far as the pollen-receptive stage, and pollen on the surface of the style of the same flower does not germinate. At maturity, however, the tip of the style splits into two sections that roll back, exposing a clean, pollen-receptive stigma to which pollen from other flowers is brought by insects. Because all of the flowers in a head do not mature simultaneously, there is assurance that pollen will be available when stigmas are mature.

One of the most unusual pollinating systems among the angiosperms is found in some members of the orchid family. Flowers of these plants are pollinated by certain species of wasps. In these plants there is an amazing relationship between the structure of the flower and the pollinating insect that must have involved a closely interconnected evolution of both plant and animal. The central pollen- and ovule-bearing part of a flower has a structural pattern that resembles to a striking degree the female of the species of wasp involved (Fig. 5-7). A male wasp, attracted to what

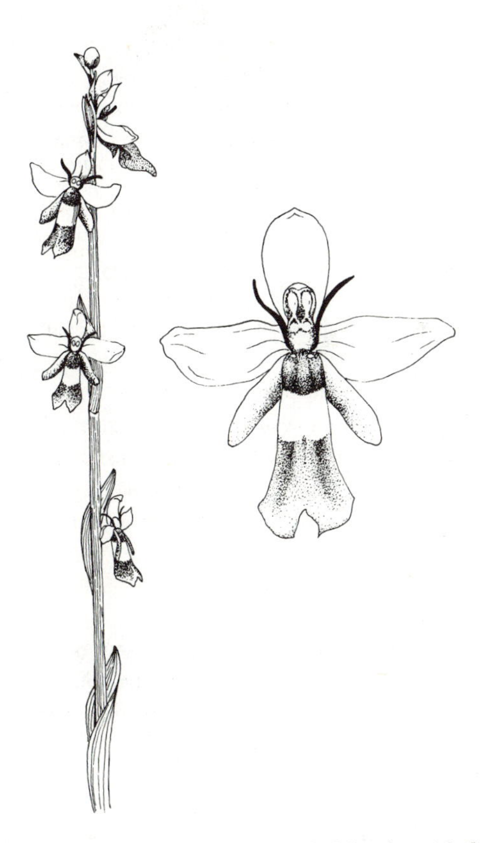

Fig. 5-7 Ophrys muscifera, *a member of the orchid family, with flowers resembling the females of a certain species of wasp. (Adapted from B. J. D. Meeuse,* The Story of Pollination, *The Ronald Press Company, Copyright 1961.)*

appears to be a female, lands on that part of the flower and actually attempts copulation. By so doing, it removes pollen from the anthers and transfers it to the stigma of another flower in a similar act of "pseudocopulation." In this unusual system it is obvious that flower structure and reproductive biology are perfectly adapted to this relationship with the insect pollinator. It is also clear, on the other hand, that such a specialized system would probably be completely noneffective if the wasps involved were not available. This is a fine example of an organism finely attuned to its environment and one that would suffer with even a small change.

Flies are pollinators of many kinds of flowers. While brightly colored perianth parts serve as attractions for bees in flowers pollinated by those hymenopterids, flowers pollinated by many of the dipterids have other means of attracting the pollen vector—powerful odors. These odors are often reminiscent of carrion, and many times the flowers themselves have a color closely resembling that of spoiled meat.

Some flowers have corollas that are shaped like elongated trumpets with deep throats. In these the nectar is borne at the bottom of the corolla tube and is unavailable to many kinds of insects. Butterflies may be pollinating agents in these instances because their elongated proboscises can reach the nectar; at the same time, the insect brushes against stamens and carpels, thereby effecting pollen transfer.

Flowers pollinated by the wind, as has been mentioned, are generally much reduced, especially in regard to brightly colored floral envelopes. Quite frequently, wind-pollinated flowers are extremely "streamlined," consisting solely of anthers or carpels. Stamen-bearing flowers frequently have a large number of pollen-bearing organs, and, furthermore, the anthers are often attached to the filaments so loosely that they sway in even a slight breeze. Wind shakes out the pollen grains, and the numerous stamens release an abundance of pollen. Because the transfer of pollen to a carpel of another flower is an extremely hit-or-miss affair when pollen is airborne, it is clear that the abundance of pollen produced by wind-pollinated plants increases the chance that at least some pollen will land on target. Although most pollen grains are "wasted" and will never be deposited on stigmas of flowers of the same species, chances are good that many grains will be transferred to the appropriate place. A further modification found in some wind-pollinated flowers (for example, grasses) is the presence of tufted or feathery stigmas that provide more surface area for pollen reception.

Modifications for Cross-pollination

Not only are flowers modified in a variety of ways to insure pollination, but an impressively large number have modifications to increase the chance of cross-pollination, that is, the transfer of pollen from a flower of one plant to the carpel of a flower from another plant. Mechanical features of certain flowers are such as to promote cross-pollination. One such modification found in some angiosperms is the production of stamens and styles of different lengths. One example might be a species of plant that has flowers of two kinds: one with long stamens and short styles; the other with short stamens and long styles (Fig. 5-8). An insect procuring nectar from a flower of the former type would have pollen deposited on its body in a more posterior position than it would from the latter type of flower. In visiting various plants with these flowers, pollen collected from flowers with long stamens would be brushed against long styles; pollen from short-stamened flowers nearer the head of the insect would brush principally against short styles. In this manner it is difficult for pollen from long-stamened flowers to be deposited on the short styles of the same or similar flowers. Some species of angiosperms with this structural modification have an even more elaborate system, with stamens and styles of three lengths.

A temporal difference in the maturation of stamens and carpels within a flower is another modification that discourages self-pollination and increases the chances for cross-pollination. Certain flowers produce stamens that mature before the carpels, or carpels that mature before the stamens. In that way, pollen shed by stamens of one

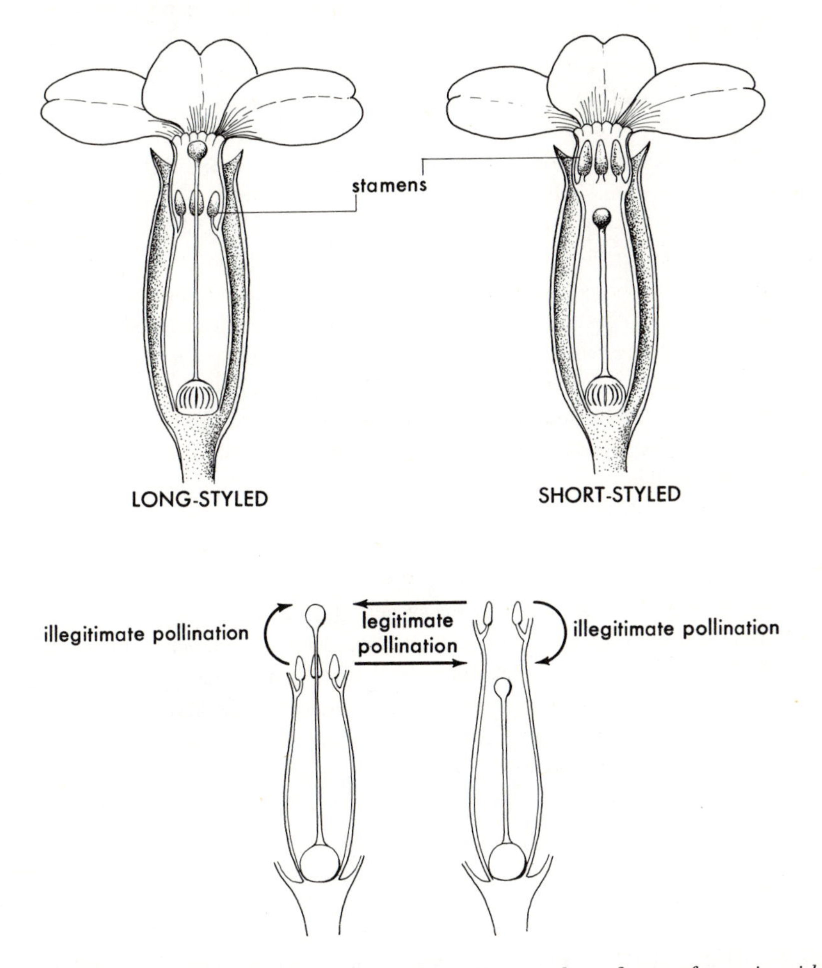

Fig 5-8 *Diagram showing the phenomenon of heterostyly, or flowers of a species with two lengths of styles and stamens. The desired course of pollen transfer is indicated. (Adapted from B. J. D. Meeuse,* The Story of Pollination, *The Ronald Press Company, Copyright 1961.)*

flower cannot grow on the stigmas of carpels in the same flower because the latter are not in a pollen-receptive stage. Conversely, when the stigma of a carpel of one flower is in the pollen-receptive condition, pollen in the same flower has been shed already or is not yet ready to be shed.

Many angiosperms have separate stamen-bearing and carpel-bearing flowers on the same plant. This separation of stems and carpels is another floral modification that does not necessarily rule out self-pollination but enhances chances for cross-pollination.

The foregoing modifications are all designed to promote cross-pollination, but it is obvious that in none of these cases is self-pollination ruled out. It is always possible for pollen of one plant to land on stigmas of flowers on the same plant. A number of angiosperms, however, are assured of cross-pollination at all times. In these there are two kinds of flowers, one stamen-bearing and the other carpel-bearing, but they are on separate plants; exclusively pollen-producing flowers are on one plant and exclusively seed-producing ones on a different plant. Here pollination can be effected only if pollen grains are transferred from one plant to another.

An unusual modification in some angiosperms is one that favors cross-pollination but permits self-pollination if pollen from another plant is not available. In tulip plants, for example, tubes from pollen grains of other plants grow faster in the style of a flower than do pollen grains from the same plant. These tubes reach the proximity of the egg sooner, and sperm from the faster-growing pollen tubes are involved in fertilization. If no pollen grains are available from other flowers, those of the same flower will be the ones that eventually produce the sperm cells.

Finally, the phenomenon of "self-incompatibility" is a feature of certain angiosperms that guarantees cross-pollination. In these plants there is a chemical barrier that prevents pollen grains from flowers of one plant from germinating on the stigmas of the same flower or of other flowers on the same plant.

Strangely enough, some plants have systems that allow only self-pollination. Although cross-pollinating systems allow for more genetic variation and more potential gene recombinations, some of which presumably will be adaptive, there are many advantages in self-pollinating systems. It is quite likely that in these instances the genetic makeup is of a combination that is selectively suitable for the individuals. Another consideration is that although the advantages of new genetic combinations resulting from cross-pollination are sacrificed, dependence on some outside source to transport pollen from one plant to another may leave many flowers unpollinated, while self-pollinating systems almost always provide guarantees of pollination. In fact, some self-pollinating flowers are so modified that the flower never opens but remains in a bud stage.

VEGETATIVE PROPAGATION

Some angiosperms have completely, or almost completely, dispensed with the sexual process. These plants may even have flowers, some of which may be brightly colored, but no seeds are formed in them. Propagation in these plants is by vegetative means. Roots, for example, may send up shoots some distance away from the stem of the plant, thereby establishing new individuals. Subsequent budding from these roots spreads the plants for some distance from the original one. Other vegetatively reproducing angiosperms may establish new plants by developing buds on the leaves that drop off.

There are obvious advantages to a system of vegetative reproduction. Here there is no need for an external agent to transfer pollen to the stigma of a carpel. The individual plant is completely equipped in itself to start new progeny. Although the beneficial effect of sexual reproduction and the accompanying recombinations of genetic characters are generally forgone, quite frequently there may be found in plants that seem to reproduce entirely by vegetative means instances of pollen successfully growing on the stigma of a carpel, with subsequent fertilization and embryo development. Those plants that reproduce by both sexual and vegetative means have a very reliable means of reproduction (vegetative) and genetic recombination of sufficient frequency, even though it is of rare occurrence, to engender them successfully adapted to their niches. It is the rule, rather than the exception, that both sexual and vegetative means of reproduction are combined within one plant.

Among certain plants cultivated by people, however, sexual reproduction may be absent, and they are propagated by shoot cuttings, root cuttings, or by some other means. Bananas are examples of plants that seldom produce viable seed. These cultivated plants would be badly suited to their environments except that humans create a tolerable set of conditions for them. When the environment changes for some reason (for example, insect pests are introduced), these plants are in danger of extinction. The plant breeder, however, may have taken advantage of the rare instances of seed production and have available new kinds of strains that are much better suited to the new environment. Or the breeder may again modify the environment (the insects may be eliminated), and the vegetatively reproducing plants will continue to thrive.

The whole story of evolution of reproductive mechanisms in the angiosperms, especially with reference to pollinating systems, is a most intriguing one. It is a story involving a close relationship of floral modification with pollen vector evolution, together with environmental change. To be adaptively suitable, any mutation in floral structure must coincide with a corresponding change in insect biology, both of which collectively must be adjustable to the environment surrounding them.

ORIGIN OF HERBACEOUS ANGIOSPERMS

From the fossil record we are able to conclude that the first vascular plants that appeared on the land in the middle of the Paleozoic Era were relatively small forms, herbaceous in structure and appearance and with no secondary development of vascular tissue. Later forms that evolved from these earliest types became progressively larger, until massive structural patterns had evolved by the

Carboniferous Period. Treelike forms persisted for the rest of geologic time and are present today. However, there were persistent herbaceous types that coexisted with the arborescent giants, and some of the present-day descendents of those ancient herbaceous members are little changed from their ancestors. The extant genus of club mosses, *Lycopodium,* for example, is almost identical with its Carboniferous counterpart, *Lycopodites.*

The fossil record also tells us that the first recognizable angiosperms in the Mesozoic Era were trees. Leaves and pollen of arborescent genera of angiosperms are quite abundant, especially in strata from the Cretaceous onward. These treelike early angiosperms would suggest that the forms from which they evolved must have been arborescent as well. At least it is quite likely that there was secondary xylem and phloem in the stems of the angiosperm progenitors. At the present time, on the other hand, there is an abundance of herbaceous angiosperms, which make up most of the ground cover on the earth today. Grasses, weeds, and the like are extremely successful plants that are probably derivatives from the earlier, arborescent angiosperms.

If we read the evolutionary history of plants correctly, then, there are actually two types of herbaceous plants: One is the persistent, primitively herbaceous form; the other is the derived type. If modern angiosperm herbs are forms that came from trees, what are the evolutionary steps involved, and why are herbaceous plants so successful?

The origin of herbaceous stem types from woody plants is not completely understood. There are conflicting views as to the evolutionary sequences involved, a probable reason being that conceivably there was more than one evolutionary pathway.

Woody Stem Anatomy At this point a brief discussion of the ontogeny of certain dicotyledonous woody stems will facilitate understanding of the subsequent treatment of phylogenetic modifications of such a woody stem in the development of an herbaceous type of plant body. An angiosperm stem grows from the tip where new cells are being formed by an active area called the *apical meristem* (Fig. 5-9). Cells in this region are small, isodiametric, and thin-walled, and undergo mitoses at a relatively rapid rate. As the apical meristem grows on, the cells formed by it begin their processes of differentiation into mature tissues. The first noticeable change is in clusters of cells arranged in a discontinuous cylinder. Here the cells are grouped into strands, and are narrower and longer than those in the surrounding tissue. These strands of slender, elongated cells are called *procambium, procambial strands,* or *provascular tissue* (Fig. 5-10A); upon maturing they will produce the first conducting tissue in the form of vascular strands. As the region of the stem below the apical meristem develops, the outer cells of the stem tip become differentiated into an epidermis, one cell in thickness. The innermost cells of the procambial strands mature into xylem elements, the primary xylem (Fig. 5-10B). The outer part of the procambial strands matures into primary phloem

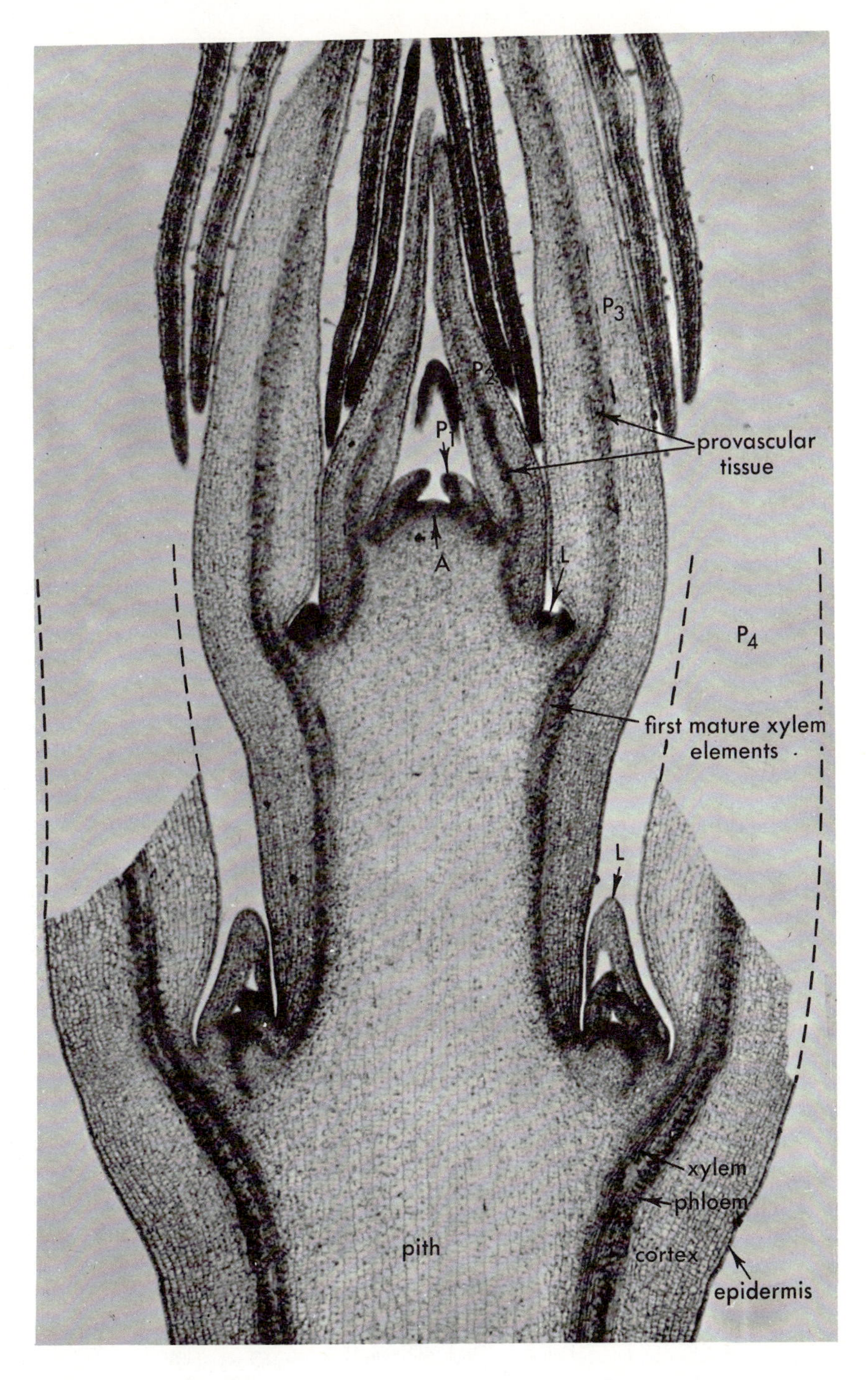

Fig. 5-9 *Longitudinal section of the shoot tip of lilac* (Syringa vulgaris). *A—apical meristem; P_1, P_2, etc.—leaf primordia; L—primordium of bud in angle between leaf primordium and stem.* ×*43.* (*From* The Living Plant, Second Edition, *by Peter Martin Ray. Copyright © 1963, 1972 by Holt, Rinehart and Winston. Reprinted by permission of Holt, Rinehart and Winston.*)

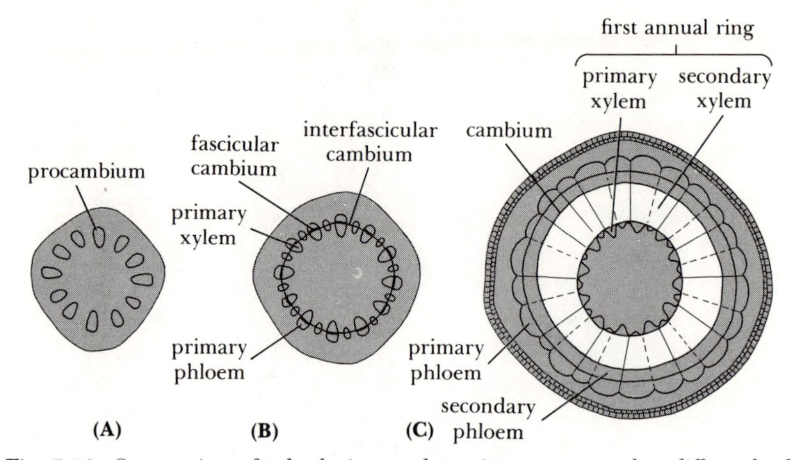

Fig. 5-10 *Cross sections of a developing woody angiosperm stem at three different levels. (A) Section just below the apical meristem, showing a cylinder of procambial strands. (B) Section of an older level, at which the inner part of the procambium has matured into primary xylem and the outer part into primary phloem, with fascicular cambium between. (C) Section of the stem, with secondary vascular tissues produced by the vascular cambium. Cork cells are formed near the periphery of the stem. (From* Botany, Fifth Edition, *by C. L. Wilson, W. E. Loomis, and T. A. Steeves. Copyright 1952, © 1957, 1962, 1967, 1971 by Holt, Rinehart and Winston. Reprinted by permission of Holt, Rinehart and Winston.)*

elements. Between the primary xylem and primary phloem is a persistent layer of procambial cells that do not differentiate but retain their ability to divide and produce more xylem cells on their inner faces and phloem cells on their outer faces. This layer of meristematic tissue, the vascular cambium, is responsible for the production of secondary vascular tissues. The part of the vascular cambium between the primary xylem and primary phloem is called a fascicular cambium. Later in ontogeny the tissue between adjacent vascular bundles becomes rejuvenated and begins to divide so that the vascular cambium becomes a continuous cylinder. That part of the cambium that originates between the original vascular strands is called the interfascicular cambium (Fig. 5-9B). Secondary xylem and phloem are then produced between the original strands as well as within them, and older stems have a continuous cylinder of secondary vascular tissues (Figs. 5-10C, 5-11A, 5-12A).

There are some variations of this pattern in certain angiosperms, for example, the production of a continuous cylinder of procambium rather than the arrangement of the procambial cells in strands.

Obviously, the continued growth in diameter of the stem due to the addition of secondary vascular tissue places strains on the outer tissues such as cortex and epidermis. In fact, the epidermis and outer cortex soon rupture because of this pressure from within. In the outer cortex some of the cells become rejuvenated, begin to divide, and produce cork cells on their outer faces (Fig. 5-10C). These cork cells then become the outer, limiting layer of the stem. Their function is primarily to prevent excessive water evaporation from the surface of the stem.

During the lifetime of most trees, therefore, there is an annual increment of growth in the stem that is the result of the addition of secondary xylem, secondary phloem, and cork cells. The tree may live for many years, and it may attain tremendous height and girth. In fact, the principal bulk of a tree is the huge cylinder of secondary wood, the function of which is to transport water and dissolved minerals to more distal regions of the tree. Of the entire volume of wood, only a very small percentage (the outermost part, or *sapwood*) is involved in actual conduction. The wood toward the center of the trunk, often recognizable in a cross section of a log by the darker color, is actually nonfunctional.

A tremendous amount of energy and organic compounds are tied up in cork formation and in the production of the woody cylinder in a tree. Because the wood itself really contributes very little to the prime function of the plant, the production of seeds, in many respects it is a very inefficient system, with a great deal of wasted

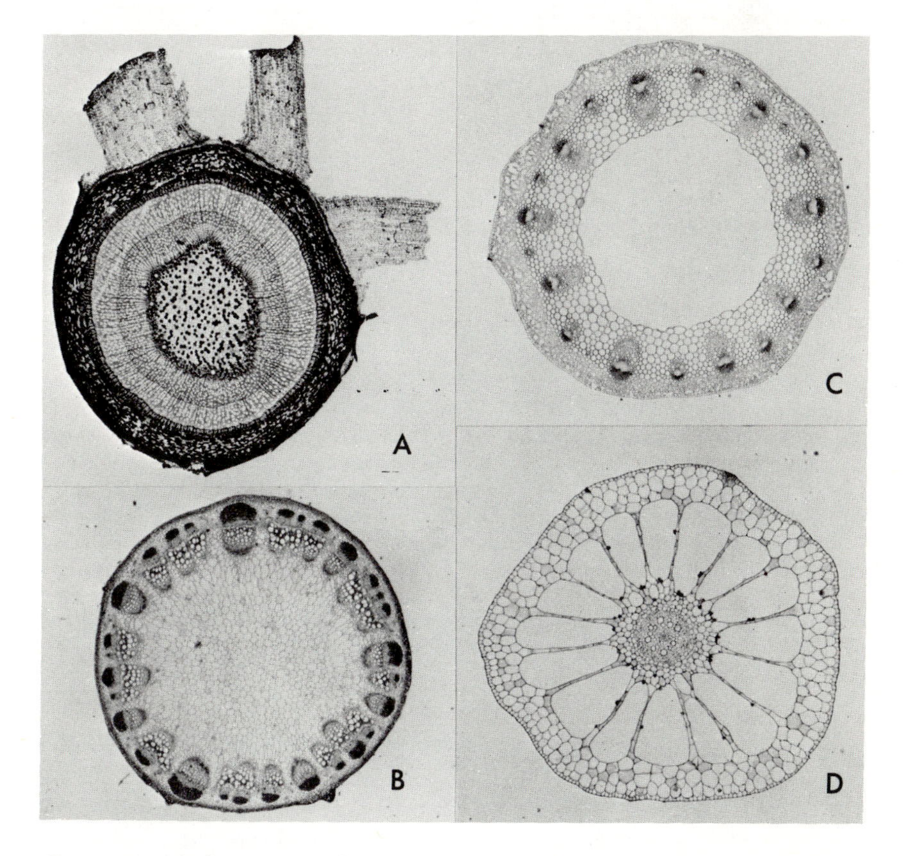

Fig. 5-11 *(A) Cross section of a young stem of sweet gum* (Liquidambar styraciflua), *showing the nature of a typical woody stem. ×20. (B) Transverse section of a sunflower stem* (Helianthus *sp.*). *Secondary vascular tissues are present but are confined to separate strands. ×14. (C) Cross section of an herbaceous plant, buttercup* (Ranunculus *sp.*). *Vascular bundles are arranged in a cylinder and have no secondary tissues. The pith is hollow. ×14. (D) Cross section of the stem of parrot's feather* (Myriophyllum), *an aquatic angiosperm. Note the large air spaces in the cortex. ×25.*

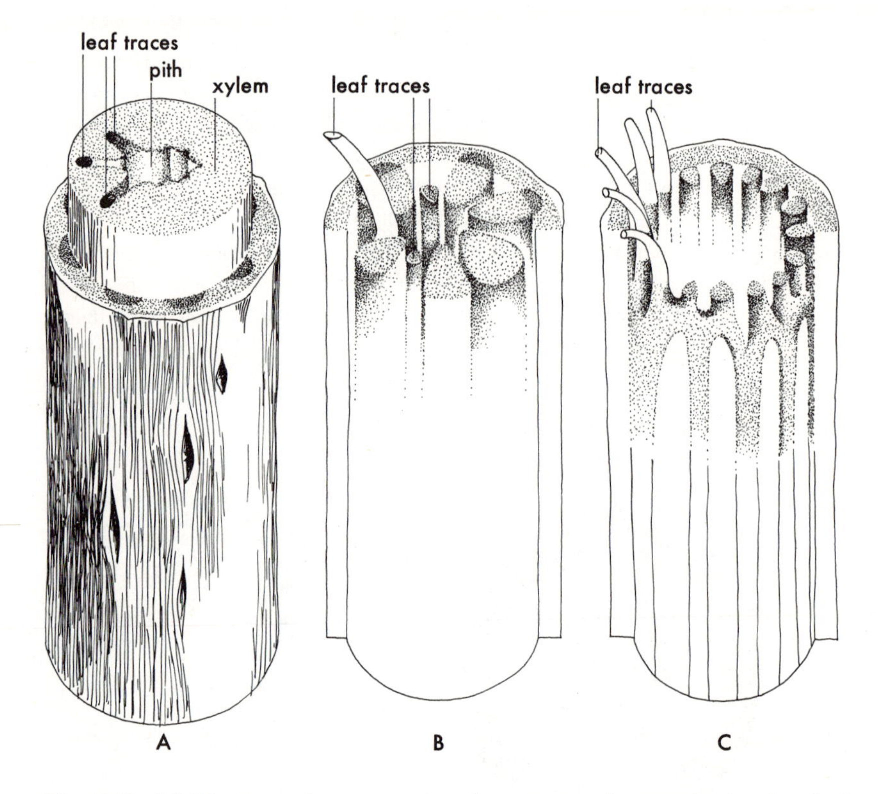

Fig. 5-12 (*A*) *Diagrammatic representation of a young woody stem, showing three leaf traces within the secondary wood and parenchyma cells associated with them on their outward course toward a leaf* (*compare with Fig. 5-11A*). (*B*) *Diagrammatic representation of an herbaceous plant with secondary vascular tissues. An increased amount of parenchyma associated with leaf traces may have been the cause of dissection of the woody cylinder* (*compare with Fig. 5-11B*). (*C*) *An herbaceous stem, diagrammatic, showing slender vascular bundles arranged in a discontinuous cylinder* (*compare with Fig. 5-11C*). (*Redrawn from* General Botany for Colleges *by Ray E. Torrey. Copyright 1922, 1925, 1932 by R. E. Torrey. Reproduced by permission of Prentice-Hall, Inc.*)

material and energy. However, another consideration in evaluating the design and efficiency of a tree, especially in temperate and cold temperate regions, is the fact that the tree must somehow make provisions for lasting through many winters. During the winter when the soil is frozen, water is practically unavailable to the tree; it is living in a virtual desert. Leaves are generally dropped from trees as a means of further cutting down water loss in the form of vapor. A thin layer of cork cells seals over the leaf scar, thus preventing water from escaping through those areas where leaves were attached. Certain trees, of course, some of them angiosperms, retain their leaves during the winter. In these plants there is often some means of cutting down excessive loss of water by structural adaptations (for example, stomata may be sunken into pits in the leaf surface, thus preventing too great a diffusion gradient from the leaf surface to the outside atmosphere, or extremely thick cuticles may be present), but these plants are really in the minority with respect to the woody plants that survive through the winters some distance from the equator. In the spring a whole new set of leaves expands, most of them within an extremely short period of time. Again, a great deal of food and energy are expended in this rapid leaf production.

These activities of woody plants—that is, the building up of an extensive woody skeleton and cork mantle and the repeated production of leaves every spring—really contribute nothing directly to efficiency in flower, fruit, and seed production. Many of the adaptations are for survival in the desiccating winter and not for seed formation. If one were given the task of designing a plant that would be more efficient, he would have to find some way of minimizing the energy expended in winter survival. Herbaceous angiosperms have just such modifications, and it is possible to demonstrate structural modifications in stems of angiosperms showing transitions from woody, treelike forms to herbaceous annuals.

Possible Origins of Herbaceous Forms

One theory concerning possible origin of herbaceous forms suggests that simply by a progressive decrease in the amount of secondary xylem and phloem, a stem structure such as that in an herb would ultimately result. This decrease would be a phylogenetic one and would be expressed ontogenetically by a failure of the cambium to produce much secondary xylem and phloem. In most herbs, the xylem and phloem are confined to strands within the stem that are surrounded by living *parenchyma cells* (Figs. 5-11C, 5-12C). If a woody stem stopped its ontogenetic development of tissues at a stage such as the one found fairly near the tip, the conducting tissues would be confined to separate strands. Cambium between adjacent bundles would never develop, and cambial activity within the bundle would be slight or, in extreme herbs, completely absent.

The second theory is considerably more complex, and involves consideration of detailed anatomical features. In woody plants, associated with a vascular trace that arises from the stele of the stem and enters a leaf (leaf trace) is an accompanying mass of thin-walled parenchyma cells (Fig. 5-12A). In fact, in a cross section of a young woody stem this parenchymatous tissue is quite obvious and interrupts the secondary xylem and phloem. This theory suggests that a progressively greater elaboration of these so-called "foliar rays" (foliar because of their association with leaf traces) serves to divide the cylinder of vascular tissue into separate wedge-shaped portions as seen in cross section (Fig. 5-12B). A subsequent reduction in the amount of cambial activity would result in a typical herbaceous stem configuration (Fig. 5-12C).

Perhaps both theories apply to different angiosperms, and it would be incorrect to conclude that only one of them provides the proper answer. Whatever the actual steps, the end result is the same, and in both cases a decrease in the amount of cambial activity is involved. Many herbs still retain cambial growth, and indeed there may be a complete cylinder of vascular tissue in the stem produced by a continuous cambium. Other herbs have the addition of secondary xylem and phloem confined to the bundles (Figs. 5-11B, 5-12B). Still others lack any secondary vascular tissue. In such forms the cambium layer may be recognizable, but apparently under normal circumstances it is completely inactive.

HERBACEOUS PLANTS AND THE ENVIRONMENT

The foregoing paragraphs explain the possible series of evolutionary modifications in the origin of herbaceous angiosperm stems, but these anatomical changes tell nothing of the biological relationships between the plants and the temperate and cold temperate environments. In many ways a tree is quite well-suited for winter conditions, whereas the stem of an herb is too delicate to survive cold and desiccating climates.

One modification that is well adapted to winter conditions in certain herbaceous plants is the production of underground organs that allow the plants to eliminate the aerial systems during winter months and to provide stored food and buds for growth during the following spring with the advent of suitable growing conditions. Underground stems, or *rhizomes,* are familiar to all who have tried to dig up certain kinds of weedy grasses. Extending outward from a parent plant are underground branches that send up new shoots some distance from the plant. By this means these grasses spread over great distances, and when the aerial portions of the plant die back in autumn, the rhizomes remain underground in a dormant condition until spring. A further modification involves more efficient food storing organs, which are actually extremely fleshy and swollen rhizomes such as those found in iris (Fig. 5-13). These structures, though resembling roots, can be shown to be stems because they have conspicuous nodes and internodes and a terminal bud.

A still greater storage volume is found in a *tuber,* also a modified underground stem. The edible part of the potato plant is just such a modified stem or tuber. In fact, this tuber is an extremely effective propagating organ. The "eyes" on a potato

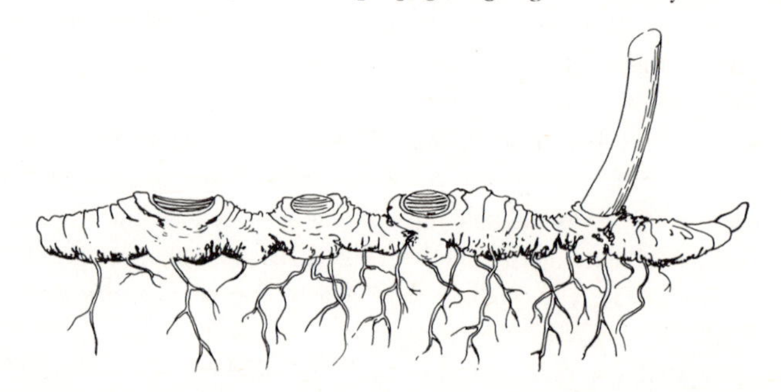

Fig. 5-13 *Rhizome (underground stem) of Solomon's seal* (Polygonatum biflorum). (*Redrawn from* General Botany for Colleges *by Ray E. Torrey. Copyright 1922, 1925, 1932 by R. E. Torrey. Reproduced by permission of Prentice-Hall, Inc.*)

are actually small buds, and it is these buds that sprout when potatoes are kept in storage too long. A farmer wishing to plant a potato crop does not use seeds but simply cuts numerous tubers into pieces, being sure that at least one "eye" is on each piece, and puts them into the ground. In the case of the potato this is of extreme importance to the farmer because potatoes do not set seed too easily. Furthermore, a good genetic stock of potatoes can be perpetuated with no difficulty by means of vegetative propagation with tubers.

By vegetative propagation these herbaceous stems with underground perennating organs can send up new shoots annually in the spring. These shoots generally do not attain great heights, nor do they have extensive secondary tissues. Flowers are formed and pollinated in one growing season, and seeds are produced before the advent of adverse conditions late in the year. After seeds are shed, the aerial part of the plant dies down, and the underground rhizome or tuber serves to carry the plant through another winter. It is obvious that in these instances there is considerable "streamlining" of the stem, with the elimination of bulky vascular tissues that contribute so little directly to the reproductive processes.

Not all herbaceous stems have underground perennating organs, however. In the most extreme herbs a seed germinates in the spring, and a new plant develops and produces flowers, fruits, and seeds. The seeds are dispersed and the plant dies, no part surviving the winter except the seeds that remain dormant during the cold months and sprout in the following spring. In other words, the entire life history of the plant is geared to the production of seeds. Practically all of the products of the photosynthetic process are diverted to fruit and seed production, with none going to special vegetative perennating organs.

Herbs as Invaders There are some obvious advantages to such a streamlined and rapid life cycle. When an area is cleared for some reason, herbaceous weed species are highly successful as primary invaders. Seeds that may have been dormant in the soil for some time or that may be blown in from some other region germinate in the open plot. Those plants that can produce a whole generation in one season are at an obvious advantage. Rapid growth processes enable them to produce seeds that are again distributed throughout the disturbed area.

Other places where herbs are successful inhabitants are areas with extremes in environment. Herbs predominate in tundras, for example, where their aerial parts die back annually.

Although the herb is an extremely successful invader, it is still not the most successful type of angiosperm plant in all areas. That this is so is demonstrated by the fact that very often the climax type of vegetation in a given region, once covered exclusively by herbaceous plants, is a forest with trees. On the other hand, a good number of herbaceous plants have seeds that are surprisingly long-lived. These seeds are quick to germinate when conditions are right and when competition from other types of plants is less severe.

A few types of herbs are successful in climax forests in spite of the dominant trees. These are often plants with underground perennating organs that send up aerial shoots early in the spring. Most of these forest herbs produce an aerial stem with flowers before many of the leaves on the trees above have developed. At this time there is still plenty of sunlight for sufficient photosynthetic activity to produce enough food material for flower and fruit development and for storage for the following year. After seed production the aerial part of the plant dies back, and only in the following spring is activity renewed. These plants occupy a niche in the climax forest that provides a short period of time during which the life cycle can be completed.

Finally, a few plants may be mentioned that have a biennial habit. During the first growing season the aerial part of the plant carries on vegetative activities only, with the photosynthates moving into underground structures, often fleshy roots, where they are stored. In the following year a new aerial part sprouts from the fleshy underground organ and produces flowers and fruits. In these plants, too, no part of the plant remains above the ground during the winter.

ENVIRONMENTAL INFLUENCES ON PLANT STRUCTURE

Different kinds of extreme environments have various kinds of profound effects on the structure of plants growing in them. These structural modifications may be regarded as adaptations allowing the plants to survive in these regions. Furthermore, often the same kinds of structural adaptations as responses to a given type of environment are found in plants of unrelated families. Thus the plants growing in such a region, though looking quite similar, are not actually related. Such a phenomenon, whereby similar structural features are attained by various groups of plants independently, is called *convergent evolution.* So closely may these unrelated plants resemble each other that it is sometimes difficult to distinguish the family without careful observation of certain features. One of the most frequently cited examples is that of the close resemblance of members of the cactus family (Cactaceae) and certain members of the spurge family (Euphorbiaceae; see Fig. 5-14). The only quick way to distinguish these families when only the vegetative aspects are present is to make an incision into the stem of each; a milky liquid will ooze out of the spurge representative.

Xerophytes

Plants growing in warm, dry regions have characteristic structural modifications. In these plants the most important features to be considered are those involved with water conservation. Plants growing in regions that are typically dry, at least for a significant part of the year, are called *xerophytes.* These plants generally have some means of survival during times of little or no water availability, but the means are not necessarily the same for all plants.

Fig. 5-14 *Convergent evolution as illustrated by a member of the spurge family* (left) *and a member of the cactus family* (right). (*With permission of the Missouri Botanical Garden.*)

The most common type of structural modification involves an extremely thick cuticular layer at the surfaces of aerial parts. This waxy layer cuts down water loss through the epidermis when stomata are closed. These xerophytes with thick cuticles still photosynthesize, however, and there are times when stomata are open. Stomata, naturally, are potential sites of water vapor loss. Most successful xerophytes have stomata depressed into the surface of the organ on which they are borne, often in pits, sometimes in furrows (Fig. 5-15A; the figure used is not of an angiosperm, but serves to demonstrate the principle involved). These depressions serve as pockets to trap water vapor, making the diffusion gradient between the inside of the organ and the outside atmosphere less pronounced. As a result, air, with its constituent carbon dioxide, can still enter the plant, while excess water vapor loss is prevented. The same principle applies to many nondeciduous plants growing in places where there are cold winters. Excessive water loss during the months in which water is not available to the plants is prevented by sunken stomata. A further modification in plants with sunken stomata is the presence of hairs in stomatal regions or over the entire leaf surface. These intertwined hairs also serve to provide many pockets of water vapor, again maintaining a low diffusion gradient between the inside of the plant and the atmosphere.

Many grasses have stomata only on the upper (adaxial) sides of the leaves, and they are arranged within longitudinal furrows (Fig. 5-15B). The furrows serve to conserve water vapor. A further effective means of maintaining internal water concentrations is by the rolling of the leaf in hot and dry weather. The chamber

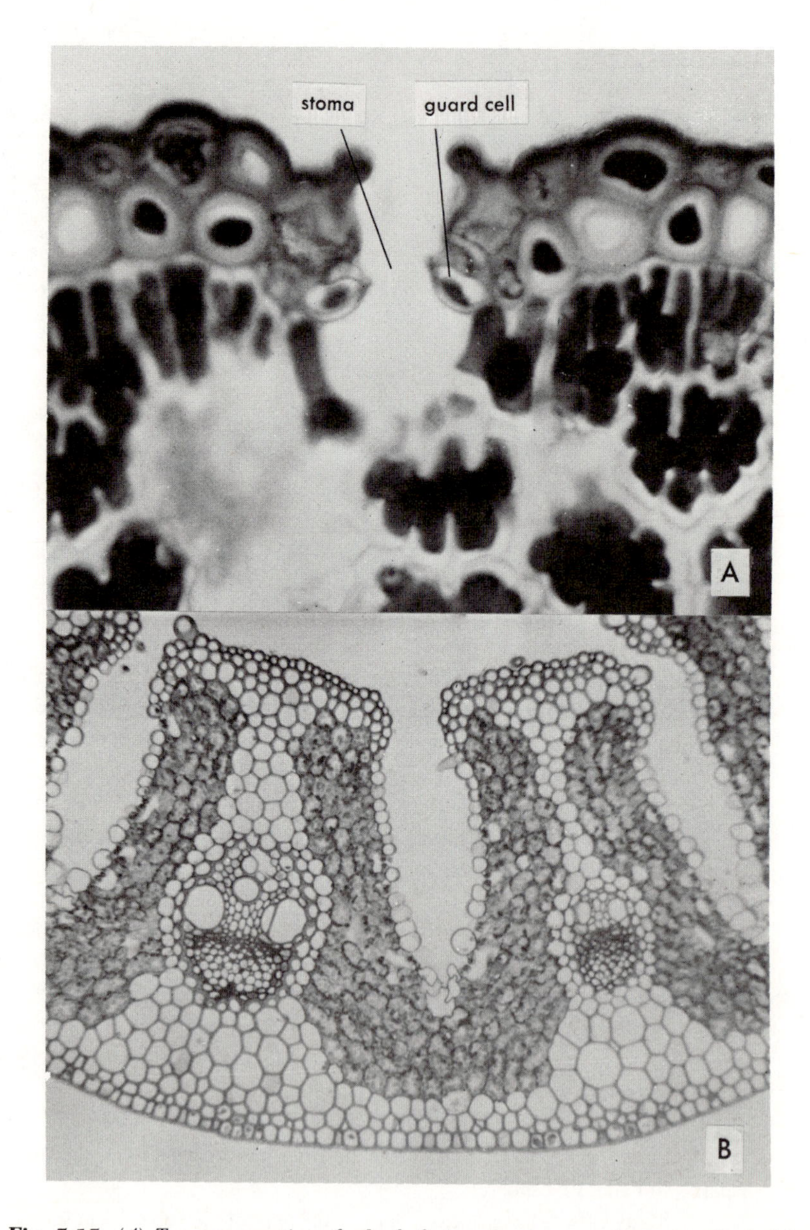

Fig. 5-15 *(A) Transverse section of a leaf of pine, showing a sunken stoma flanked by guard cells.* ×*650. (B) Cross section of part of a leaf of* Ammophila, *a grass, with longitudinal furrows on the upper (concave) face. Stomata are arranged in these furrows.* ×*180.*

formed within the upper rolled leaf surface accumulates water vapor from the leaf surface and maintains a low diffusion gradient between the interior of the leaf and the air surrounding the leaf.

Some xerophytes have means of storing necessary water for metabolic processes during unfavorable dry periods. Plants such as fleshy cacti retain great quantities of water within the plant tissues. This water, made available during occasional periods of rainfall, is taken in by the plant and stored in the conspicuously fleshy tissues of the stem.

A number of plants that grow in normally dry environments are able to complete an extremely rapid life cycle. These plants may be dormant in the form of seeds during months of dry weather. As soon as rains come, the seeds germinate, and the plants grow, mature, and produce flowers and seeds—all before the short wet season is terminated by hot, dry months.

These are just a few examples of ways in which plants are modified for survival in dry climates. Plants from different, even unrelated, families may become modified in one or more of these ways, with the end result being an assemblage of plants looking quite similar because they have the same structural and functional adaptations. However, this similarity is only coincidental. Obviously, the modifications have proved through the ages to be suitable for survival, as attested to by the fact that so many kinds of flowering plants have assumed these similar structural characteristics.

Hydrophytes

Hydrophytes, or plants that live in aqueous environments, also show certain structural features that are found in representatives of a number of families in the same environment. Typical among plants living submerged or partially submerged are conspicuous air chambers in tissues of stems, roots, and petioles (Fig. 5-11D). Some of these plants that grow partially submerged have leaves of more than one structural type. A leaf growing under the surface of the water quite commonly is dissected into filiform segments. Entire and undissected leaves are attached to the same stem at higher levels above the water. Anatomically, plants living in aqueous environments are quite reduced. The vascular system is generally less well-developed than in plants in drier habitats (water and mineral conduction is less of a problem in the water), and very often the xylem lacks vessels. Indications are that vessels are absent not because the wood is a primitive type, but because they have been lost secondarily. The full adaptive significance of many of the modifications in hydrophytes is little understood.

Experimental Studies on Environmental Effects

A number of significant experimental studies have been carried out on plants in attempts to determine the effects of environment on structural characteristics. Plants of the same genetic constitution were planted in gardens in a number of stations varying in latitude and in altitude, and observations were carried out on the structure assumed by the progeny in these localities. Immediately it became obvious that the environment plays a considerable role as an influence in the expression of the genotype. Thus the "typical" structure of a given plant depends not only on its genetic makeup, but also on the effect of the environment on the ways in which these genes express themselves.

SEED AND FRUIT DISPERSAL MECHANISMS

The success of flowering plants in their various environments depends largely on their means of propagating themselves. A given individual may be highly successful in its own niche at a given time, but there must be a means of continuing the line after the death of the individual. It is essential, therefore, for seeds to be distributed at the correct time, in the right places, and in the proper amounts. Within the angiosperms are a great variety of seed dispersal mechanisms. A given dispersal mechanism may be found in a number of unrelated families, although certain means of dispersal may be dependent upon some peculiar inherent structural characteristic of a family.

Some of these means of seed distribution depend upon other agents for the transporting of seeds. Many animals serve as vectors of seeds and fruits to places quite removed from the parent plant. One of the most common modifications in seeds and fruits distributed by animals involves a fleshy outer seed coat or edible fruit coat and a harder inner seed or fruit coat. An animal will eat the seed or fruit, digesting the softer, outer coat, passing the hard inner portion through its intestinal tract, and depositing it in some other place. The seed of magnolia, for example, has a brilliant fleshy outer part attractive to birds that eat these seeds and eliminate the harder, indigestible inner part. Furthermore, these seeds dangle on slender, flexible stalks from their opened dried fruits, and the movement of these pendant seeds in the wind serves as a further means of attracting birds. Peaches and cherries (members of the rose family) are examples of fruits that have palatable outer fruit walls, with the inner part of the fruit (the stone or pit) that contains a seed being left undigested in a place often some distance from the plant that bore it. Naturally, many of the places where the seeds are deposited are unsuitable for maintaining growth, and many of the seeds fail to germinate. Others, however, are left in places that induce germination and support the plant after it becomes established.

Some seeds and fruits are distributed by animals to which these seeds and fruits adhere by various means. Most readers have probably walked through woods or fields in the autumn only to emerge with a covering of a number of kinds of seeds or fruits adhering to clothing. Some of these seeds or fruits have many short, recurved, stiff hairs that catch onto clothing or animal fur (beggar's ticks or tick trefoils of the genus *Desmodium* of the legume family). Others have elongated processes, possibly with stiff hairs directed backward (beggar's ticks of the composite family; see Fig. 5-16B). There are many more kinds of seeds and fruits that adhere to animal dispersal agents. Many of the mechanisms involve simple hairs, although others have

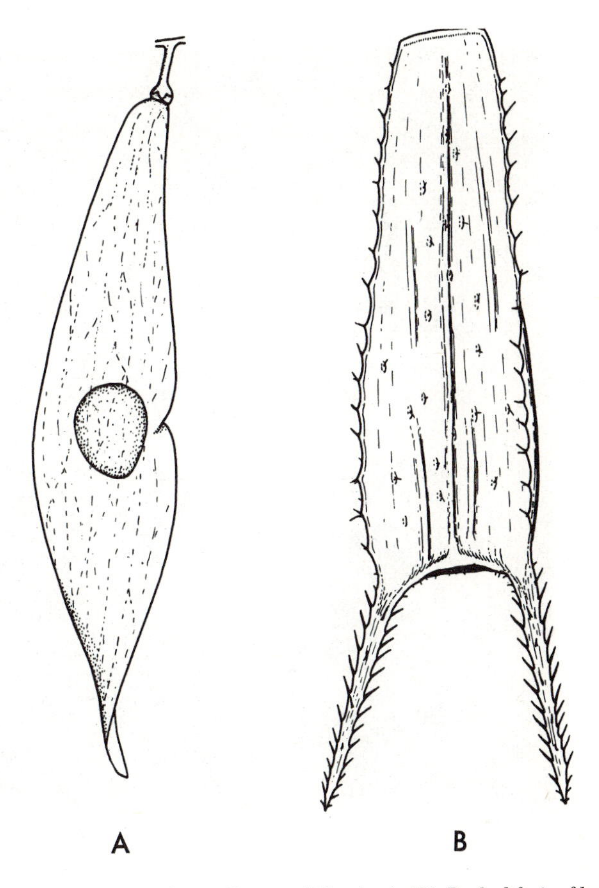

Fig. 5-16 (*A*) *Winged fruit of tree of heaven* (Ailanthus). (*B*) *Barbed fruit of beggar's tick* (Bidens). (*From* Botany, Fifth Edition, *by C. L. Wilson, W. E. Loomis, and T. A. Steeves. Copyright 1952, © 1957, 1962, 1967, 1971 by Holt, Rinehart and Winston. Reprinted by permission of Holt, Rinehart and Winston.*)

more pronounced stiff processes from the seed or fruit wall. Still others involve a modification of reduced leaves surrounding the fruit (burdock, a member of the composite family).

Wind is an important agent in the transport of fruits and seeds. The walls of fruits or the coats of seeds are so modified in these situations as to provide additional surface area for wind currents. Maple keys and ash fruits have extensions of the fruit wall to provide a kind of propellerlike structure that allows the fruits to be carried some distance by the wind (Fig. 5-16A). Seeds of catalpa have extensions on their seed coats that provide a buoyancy mechanism in air currents. Milkweed seeds, which are small, flat, brown structures, have a plume of silky hairs that carries these seeds for miles in even a light breeze (Fig. 5-17). Small fruits of other plants (for example, asters and dandelions, both in the composite family) are also plumed with a tuft of hairlike structures. These small fruits, each with only one seed within,

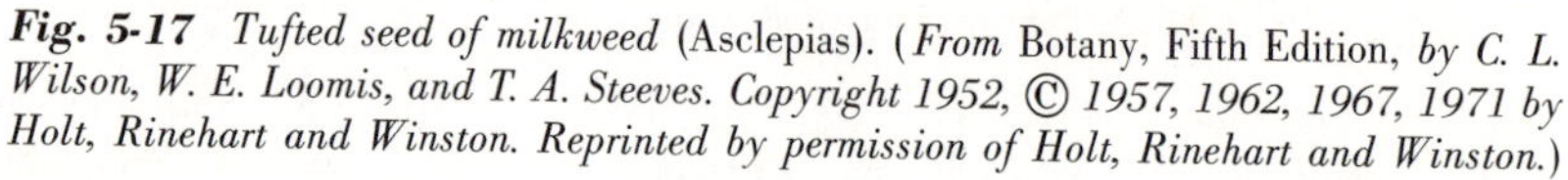

Fig. 5-17 *Tufted seed of milkweed* (Asclepias). (*From* Botany, Fifth Edition, *by C. L. Wilson, W. E. Loomis, and T. A. Steeves. Copyright 1952, © 1957, 1962, 1967, 1971 by Holt, Rinehart and Winston. Reprinted by permission of Holt, Rinehart and Winston.*)

function as seeds, with both the fruit coat and seed coat splitting at the time of germination. Other airborne seeds are relatively unmodified structurally; they are simply tiny structures that are extremely light and are carried considerable distances by wind. Seeds of orchids, for example, have embryos consisting of only a few cells. These are produced in fruits of orchids in tremendous quantities, and are blown for quite some distance. Obviously only a minute percentage germinate and thrive, but the quantities produced guarantee that some will land in favorable places.

Waterborne fruits and seeds are thought to have been instrumental in the population of islands with certain kinds of plants. In the Pacific Ocean, for example, the floras of widely separated islands may be quite similar, a probable reason being the washing-in of fruits and seeds by ocean currents. Birds, too, probably play an important part in long-distance transport.

Examples cited above are those in which the plants are dependent upon other agents for the distribution of seeds. However, many plants have structural modifications suited for forcibly ejecting seeds. A weedlike vine, the "squirting cucumber" (*Ecballium*) has a fleshy, ellipsoid, spine-covered fruit, usually with four chambers inside. At maturity enough turgor pressure is built up within the fruit to burst open one end of it, sending the seeds out from the chambers with considerable force and for quite some distance (Fig. 5-18A). Another plant, the touch-me-not (*Impatiens*) has elongated fruits, with numerous small seeds arranged within in a series of rows. As the fruit matures and begins to dry out, the wall splits suddenly and violently, tossing out the seeds forcibly (Fig. 5-18B).

The selective advantages of certain of these seed- and fruit-distributing mechanisms are obvious. On the other hand, these adaptations and many of those mentioned on earlier pages must be regarded as parts of a great number of factors that determine the total selective adaptability of the plants. Naturally, a plant that has efficient seed-distributing devices but is poorly suited for a niche where there is frequent drying out will not be better adapted than a structurally specialized xerophyte with no unusual or dramatic means of distributing seeds. To judge the adaptive value of a given feature, one must assume that all other factors are equal, which, of course, they never are.

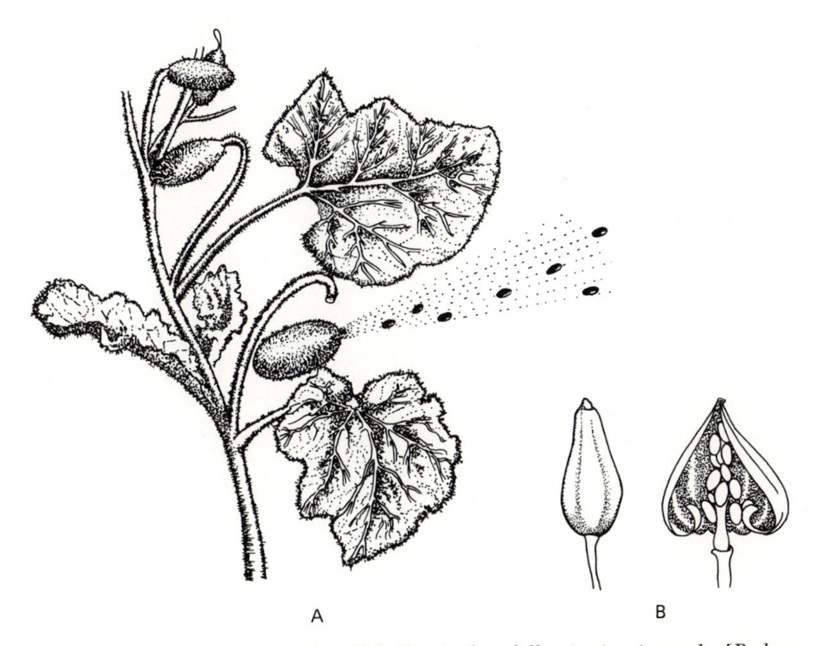

Fig. 5-18 *(A) Squirting cucumber* (Ecballium), *forcefully ejecting its seeds. [Redrawn from* The Plant Kingdom, *by William H. Brown, © Copyright, 1935, by William H. Brown. Used by permission of the publisher, Ginn and Company (Xerox Corporation.)] (B) Fruit of touch-me-not* (Impatiens), *which spreads its seeds by flinging them out when it snaps open at maturity. (Redrawn from William H. Brown,* A Textbook of General Botany, *1925, published by Ginn and Company.)*

FURTHER READING

Briggs, D., and S. M. Walters, *Plant Variation and Evolution.* New York: McGraw-Hill, 1969.

Cook, S. A., *Reproduction, Heredity, and Sexuality.* Belmont, Calif.: Wadsworth Publishing Company, 1964.

Corner, E. J. H., *The Life of Plants.* New York: Mentor Books, New American Library, Inc., 1968.

Esau, K., *Anatomy of Seed Plants.* New York: Wiley, 1960.

Grant, V., "The Fertilization of Flowers," *Scientific American,* vol. 184 (June 1951), pp. 52–56.

Meeuse, B. J. D., *The Story of Pollination.* New York: Ronald, 1961.

Ray, P. M., *The Living Plant,* 2d ed. New York: Holt, Rinehart and Winston, 1972.

Savage, J. M., *Evolution,* 3d ed. New York: Holt, Rinehart and Winston, 1977.

van der Pijl, L., *Principles of Dispersal in Higher Plants.* Berlin: Springer-Verlag, 1969.

Paleobotanical Summary

Normally, in a discussion of morphological and evolutionary principles the treatment of fossil plants is best included along with that of contemporary forms. This has been done to a large extent in earlier pages. A consideration of the fossil record by itself, however, is of considerable interest in tracing changes in plant form through periods of time. Although there are various ways of attempting to determine the mechanisms involved in plant diversification and the many interacting factors involved in inducing changes in form in plants, only a study of the record of the plant kingdom preserved in rocks can produce an overall picture of the actual changes that took place and of the times in the earth's history during which these changes occurred.

SIMPLEST PLANT FORMS

The fossil record indicates that the first recognizable plant forms occurred during the Precambrian Period, at a time approaching two billion years ago. There are indirect pieces of evidence that plant life was much older, but the ones mentioned here are all fossils with structural preservation, rather than possible organic remains. These structurally preserved plants are simple filamentous forms most closely resembling modern blue-green and green algae.

Plant life during the Precambrian seems to have consisted only of nonvascular types, most likely living in aquatic habitats. Evidence that algal forms of plant life existed during the Precambrian occurs in the form of deposits of calcium carbonate in unusual patterns that suggest deposition by living things. Rather than being deposited in stratified horizontal layers, these deposits of calcium carbonate were concentrically formed, with layer after layer being added around a central nucleus. These concentric deposits of lime resemble modern blue-green algal deposits called water biscuits. Some of the calcium carbonate deposits of Precambrian times are quite similar to those of blue-greens today; others are more closely comparable to lime-precipitating red algae.

Algal remains of other groups begin to show up in the fossil record of the Paleozoic Era, with progressively more forms known in progressively younger strata. In fact, even before the Tertiary Period, fossil remains are known of the blue-green algae, green algae (there were many lime-precipitating forms among them), golden algae, dinoflagellates, brown algae, and red algae (of which there were also many that precipitated lime).

The fungal record is less abundant, especially in earlier rocks. Many kinds of fungus filaments were preserved because they occur within tissues of vascular plants that were fossilized. There is a report of the occurrence of the genus *Albugo,* the organism causing white rust in vascular plants, in seeds of Carboniferous gymnosperms. There are also reports of a symbiotic relationship between fungi and roots of higher plants as far back as the Carboniferous Period. During the Tertiary many kinds of fungi lived parasitically on leaves of flowering plants (Fig. 6-1). A number of these are referable to genera and families of extant forms.

A number of bryophytelike plants have an old history. Certain bryophytes are recognizable because their structure closely parallels modern mosses and liverworts. These extend as far back as the Devonian Period. Other puzzling plants are assigned to the bryophytes because their remains suggest a life history at about the same degree of specialization as that found in modern bryophytes, but not quite resembling the modern forms. Late Mesozoic and Tertiary mosses and liverworts are much more abundant and quite closely allied to modern forms.

PRIMITIVE VASCULAR PLANTS

It is the vascular plant history that is most clearly recorded in the rocks. Tissues in land vascular plants are more resistant to decay than are those of nonvascular plants, and these types of plants are preserved in a variety of ways in

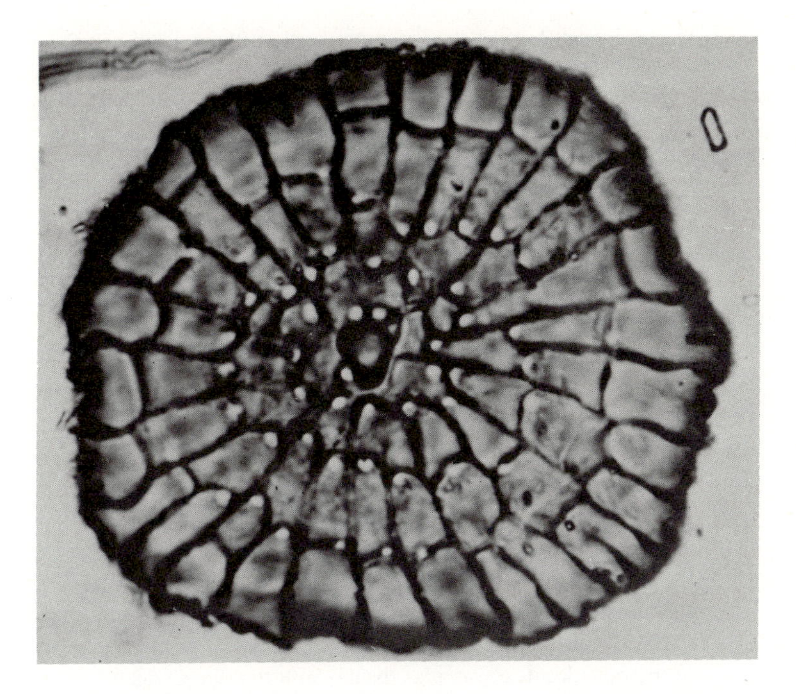

Fig. 6-1 Callimothallus pertusa, *a fungus parasitic on an Eocene angiosperm leaf. ×1500. (Photograph from a slide made available by D. L. Dilcher.)*

the earth's crust. In the paleobotanical literature are occasional reports of land vascular plants as far back as the Cambrian Period. These reports are based on spores, bits of purported vascular tissue, and occasionally on carbonaceous compressions of structures that have some resemblance to vascular plants. Many of these reports are not free from question, however, and it is generally conceded that the first undisputed vascular plants appear in the fossil record in upper Silurian rocks. In recent years, even some of the presumed late Silurian vascular plant remains were shown to have been misdated early Devonian forms. During late Silurian and early Devonian times, relatively simple vascular plants existed, and even in that early period more than one group was evident. The simplest were members of the rhyniophytes—plants that were dichotomously branched, leafless, and with sporangia borne terminally. These forms also had extremely simple vascular strands in the form of small protosteles. Coexisting with them were zosterophyllophytes, which had a somewhat similar dichotomously branched habit but with the sporangia borne along the sides of the axes. These plants were also protostelic. A third major group, the trimerophytes, showed a distinction between a prominent main axis and smaller, dichotomously branched laterals. These plants are most likely derivatives of the rhyniophytes. Appearing quite early in the fossil record are lycopods, which had closely spaced, helically arranged leaves on dichotomizing axes. Their origin appears to have been derived from specialization of the spinelike projections on the axes of certain zosterophyllophytes. When present, sporangia were generally on the upper surfaces of small leaves. Steles were typically without a pith, and they were often lobed in cross-sectional outline. Relatives of the horsetails also had dichotomously branched axes, on which were borne delicate dichotomously branched leaves in whorls. Sporangia were borne on the tips of some of these leaflike structures; the tips were generally bent back toward the axis from which they arose. Finally, fernlike plants made their appearance in the Devonian. These primitive ferns combine characteristics of the simply constructed trimerophytes with more advanced structural specialization at the tips of axes, which may be planated with leaflike structures.

Even though these several groups of vascular plants are recognizable as early as the Devonian Period, they all show certain features in common, and all have a basic stem plan composed of or derived from a dichotomizing system. The main difference involves the kinds of appendages borne on these axes. The gross similarity of body plan has led many morphologists to assume that the rhyniophyte type of plant body was the most primitive type from which other forms originated. A conflicting school suggests that because there are several groups that are distinct as far back as the Devonian Period, these groups must have had independent origins, each from a different algal ancestor.

However, when we consider the plants in their entirety, they have much in common, perhaps more than could be explained as a coincidental convergence. For example, the life histories of all of these early vascular plants are similar. All of them have the same kinds of vascular tissues, with the same evolutionary trends having occurred in the specializations of these steles. Modern counterparts, moreover, are similar chemically, with the same kinds of photosynthetic pigments and food-storage products. It must be admitted that there are strong arguments for both the monophyletic theory and the polyphyletic theory of vascular plant origin; the answer will come only when these simplest, earliest forms are more completely understood, or when more of these plants are discovered as fossils.

EVOLUTION OF SEED PLANTS

The seed plant line probably originated from some of the Devonian progymnosperms. Certain of these presumed ancestors developed extensive secondary vascular tissues, and sporangia were borne on branch systems that superficially resembled large, compound leaves. Some forms of sporangia were heterosporous, a condition necessary for the origin of seeds.

The earliest report of seedlike bodies is from the upper Devonian. These early fossil ovules are primitive in the sense that they consist essentially of megasporangia incompletely enveloped by sterile, fingerlike processes arising from the sporangial stalk at the base of the sporangium (nucellus). More advanced fossil ovules from the Mississippian and later periods have considerably more specialized integuments, consisting of a complete envelope around the megasporangium. During the Carboniferous and Permian Periods a very important group of plants, the seed ferns (often called pteridosperms) flourished; these probably had their origin from some of the relatively simpler progymnospermous plants of the late Devonian and early Mississippian that were heterosporous or that had incompletely enveloped megasporangia such as the type mentioned earlier for *Genomosperma.*

Contemporary with the seed ferns in the late Paleozoic was another group of seed plants, the conifers. These plants generally had simple or sparsely divided leaves, as opposed to the generally conspicuously compound leaves of the seed ferns. Furthermore, there was a tendency for compaction of seed-bearing structures into cones so typical of conifers at the present time.

By the end of the Paleozoic Era, therefore, a number of distinct groups of vascular plants were in existence. Still present were lycopods, represented by large arborescent forms that were especially abundant and actually dominant in the forests of the Carboniferous and Permian Periods. Horsetails also had treelike forms during the Carboniferous and Permian, and were probably second in importance only to the lycopods. Ferns were abundant, and the great number of fernlike leaf fossils during the Carboniferous Period is responsible for the use of the phrase "Age of Ferns" in reference to that geologic period. Many of the fernlike leaves, however, actually belonged to the seed ferns, gymnospermous plants, the foliage of which is often indistinguishable from that of the true ferns (Fig. 6-2). Finally, the conifers were another major group that was well established before the end of the Paleozoic Era. This list does not exhaust all the kinds of plants that were present before the onset of the Mesozoic. Many others groups of plants, most without modern representatives, coexisted with the groups mentioned above. The ones listed here, however, are the most important as far as our knowledge of forest composition during the Paleozoic Era is concerned.

Ferns, lycopods, and horsetails persisted into the Mesozoic, but that era is marked by the surgence of seed plants. In fact, the Mesozoic is frequently called the "Age of Cycads" because foliage of that group of seed plants is especially abundant. Actually, the so-called "cycads" represent more than one group of seed plants; there were relatives of the modern order Cycadales, but also quite important were plants with a similar superficial appearance, the Cycadeoidales (Bennettitales in some texts). Conifers, too, were conspicuous members of the Mesozoic forests, with most of the extant conifer families appearing for the first time during that era. Seed ferns persisted into the Jurassic Period, after which time there is no further record of them. *Ginkgo* relatives, too, had their origins in the Mesozoic, or perhaps even before (Fig. 6-3).

During the Mesozoic the lycopods and horsetails gradually dwindled in importance, with progressively fewer forms of those groups occurring in the latter part of the era. As a general rule, ferns living then resembled forms in modern families quite closely; some, however, were types not found today. Fern distribution during the Mesozoic Era was considerably more significant than at the present time.

The most important evolutionary event that occurred during the Mesozoic Era was the first appearance of angiosperms. Exactly when they appeared first is still unknown, but the Cretaceous Period is the time of their rapid spread. Angiosperm-like plants in the form of impressions resembling palm leaves are known from the

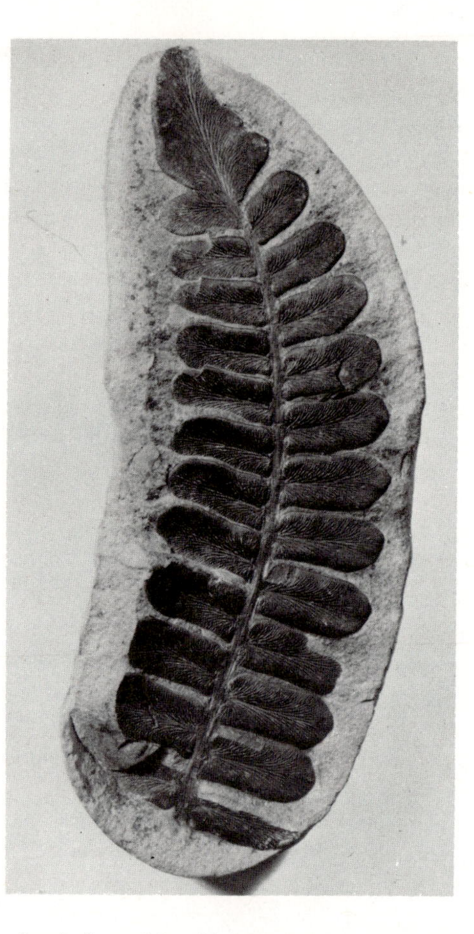

Fig. 6-2 *Part of a frond of a seed fern,* Neuropteris gigantea, *from the Pennsylvanian of Illinois. Note the resemblance to fern foliage.* $\times 0.75$. (*From* Morphology and Evolution of Fossil Plants, *by Theodore Delevoryas. Copyright © 1962 by Holt, Rinehart and Winston. Reprinted by permission of Holt, Rinehart and Winston.*)

Triassic Period. Other pieces of evidence that are still not completely validated consist of pollen grains, pieces of wood, and leaf impressions. Whatever their origin, (the group or groups from which they arose are still unknown), it is certain that their spread was spectacularly rapid. Early in the Cretaceous, angiosperm fossils were few, but they increased in number and in geographic range later in the Cretaceous, and by the end of the period they had attained an amazingly widespread distribution. In fact, by the late Cretaceous the flowering plants had become the dominant group of vascular plants on the earth.

Coinciding with the advent of flowering plants, the gymnospermous plants, especially the cycads, began to assume lesser importance, and by the end of the Cretaceous, one of the groups of seed plants, the Cycadeoidales, became extinct. Seed fern relatives had disappeared somewhat earlier. Lycopods and horsetails are

Fig. 6-3 *Jurassic ginkgophyte leaves* (Ginkgo huttoni) *from Yorkshire, England.* ×*0.7.*

inconspicuous components of floras at the advent of the Tertiary, and ferns were less important than they had been earlier in the Mesozoic.

By the Tertiary Period all of the presently known plant groups had made their appearance. There was considerable climatic fluctuation during that period, and floras of that time reflect these changes, but there was really little major evolutionary change.

Imperfect as it is, the record of plant life as represented by fossils is a significant key to our understanding of the structure and distribution of modern kinds of plants. Certainly it is not the only key, but all of our ideas concerning mechanisms of evolutionary change are valid only when they are compared with the actual recording of evolutionary events in the rocks.

FURTHER READING

Andrews, H. N., Jr., *Ancient Plants and the World They Lived In.* Ithaca, N.Y.: Comstock Publishing Associates, 1947.

Andrews, H. N., Jr., *Studies in Paleobotany.* New York: Wiley, 1961.

Arnold, C. A., *An Introduction to Paleobotany.* New York: McGraw-Hill, 1947.

Banks, H. P., *Evolution and Plants of the Past.* Belmont, Calif.: Wadsworth Publishing Company, 1970.

Delevoryas, T., *Morphology and Evolution of Fossil Plants.* New York: Holt, Rinehart and Winston, 1962.

Index